Damages on Pumps and Systems

Damages on Pumps and Systems

The Handbook for the Operation of Centrifugal Pumps

Thomas Merkle
M. Eng. Dipl.-Ing. (FH)

AMSTERDAM • BOSTON • HEIDELBERG • LONDON
NEW YORK • OXFORD • PARIS • SAN DIEGO
SAN FRANCISCO • SINGAPORE • SYDNEY • TOKYO

Elsevier
225 Wyman Street, Waltham, MA 02451, USA
The Boulevard, Langford Lane, Kidlington, Oxford, OX5 1GB, UK

Notice
No responsibility is assumed by the publisher for any injury and/or damage to persons or
property as a matter of products liability, negligence or otherwise, or from any use or
operation of any methods, products, instructions or ideas contained in the material herein

British Library Cataloguing in Publication Data
A catalogue record for this book is available from the British Library

Library of Congress Cataloging-in-Publication Data
A catalog record for this book is available from the Library of Congress

ISBN: 978-0-444-63366-8

For information on all Elsevier publications
visit our web site at store.elsevier.com

**Working together
to grow libraries in
developing countries**

www.elsevier.com • www.bookaid.org

Contents

Preface

Centrifugal pumps besides being used for transportation of pure fluids, are also very frequently used for transportation of fluids with solid components. Abrasive media drag on volutes and impellers so strongly, that considerable damages occur to the components of the pumps.

Consequently, depending on strain and composition of the medium to be pumped and the solid matters contained in it, only a low service life could be achieved and therefore being uneconomic. A wrong operation of pumps can result to cavitation and cause severe harms altogether. Under the aspects of Total Cost of Ownership (TCO) and Life Cycle Cost (LCC), this means a drastic increase of operating costs.

In this book it is described, that with suitable measures, wear up can be reduced and partly prevented. Constructive measures, preventive maintenance, optimal servicing and repair of systems or plants, can both, extend the life cycle and save expenditures.

Even the application of speed control, new technologies for overcoating and production of pump components, can also increase the profitability of pump systems considerably.

Using concrete examples, damage mechanisms and fundamental relations will be shown and evaluated. Indications for fault management as well as suggestions on measures for failure prevention and fault detection will also be analysed.

The specific wear up for example on "pumping of metal chips", will also be described in theory and practice.

Pumps which are used in the field of shaving metal processing metal devices and machines for transportation of abrasive fluids, are oblieged to completely different operating conditions rather than those for transportation of pure, clean liquids. It will be explained, that a prognostic maintenance can be handled economically.

Deep moving, theoretical fundamentals and calculations were renounced, since many technical literature is already sufficiently available on these topics.

In this english version of my book "Schäden an Pumpen und Pumpensystemen", I quite thank cordially Dr. Isack Majura for his support by tips and proof readings of the text. In this account, his suggestions were very helpful to write the book in a user friendly and easy reading shape.

This book is intended as a guide to be able to minimize or avoid damages. Indeed, it's a practical help for planners, plant engineers and operators of systems in the field of shaving metal processing, as well as areas of maintenance, servicing and repair of equipment where pumps are used. Moreover, students in the fields of mechanical engineering and process engineering, it might be a helpful and valuable support for study.

Thomas Merkle
Tübingen, January 2014

1 Introduction

Damages and wear out on pumps can have very different causes. Longtime damages appear often only after years. The type and intensity of the strain of the pump has a very decisive influence on it. Short, timed operation, cyclical or continuous operation determine the lifetime of the pump indeed. Damages which occur only shortly after going into operation, are very often due to faulty planning or mismanagement. A wrongly dimensioned pump, or the operation outside the operating point, can lead very quickly to the failure or complete breakdown of the pump.

1.1 Causes and Effects of Wear Out on Centrifugal Pumps

Very different strains can cause disturbances, wear out or also the total failure of a pump. For example these could be foreign elements in the volute, the impeller or in the pipeline. Other mistakes could be overload, wrong operation, which leads to cavitation or a faulty mechanical seal, which has the leakage as the consequence.

The transport of fluids with solid matters, which are moreover hard, causes a harmful abrasion on pump components. The reduction of the flow speed has a decreasing effect on wear out.

Delivery of special media, which are strongly alkaline, saline solution, acids or also seawater, cause more or minor corrosion effects. Also unplanned load modification without power adjustment have negative consequences.

1.1.1 Foreign Particles in the System

Solid particles like metal chips, deposits, grinding particle (sand, corundum etc.) but also components like screws, nuts or broken drills, which are very often delivered inside the fluids, could however destroy the pump. It is recommended, if solid matters with several millimeters of diameters are in the fluid and have to be pumped, a special free flow pump with an open impeller should be used (see figure 1).

There are already pumps available for processes where metal chips with a length of several millimeters must be pumped out. These pumps are equipped with a cutting device. The metal chips are already cut outside the volute to small pieces before entering into the pump. The small chopped chips could then be drawn in by the pump.

The solid particles do not lead as in the case of the closed impellers to blockage but are dragged along in the volute by the liquid to be transported out of the pump volute over the pressure filler neck again.

By operation with free flow pumps, the solid particles don't really enter the impeller, as shown in figure 2. The fluid with the solid components is drawn in by the impeller into the volute. By circulating under speed, the medium with the solid skids out again. [31]

Figure 1 Free flow pump with confining chamber

It can be problematic at a solid matter quota of more than 10%, though. This, however, depends on the granulation or the size of the solid matters. Often, only an empirical ascertaining gives a solution.

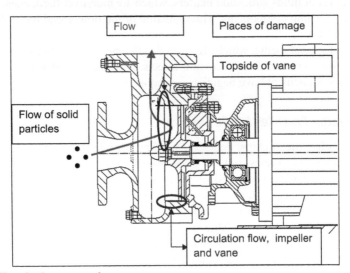

Figure 2 Flow in the pump volute

A longer-term operation with abrasive media leads however to wear out appearances, though.

Faulty spots arise, depending on the type of the solid matter, differently strong both on the vane top side of the impeller and on vane circulation side.

It is still the simplest opportunity, to replace the impeller after the damage. A complete exchange of the pump is often inevitable at greater damages to the volute.

It is recommended preventively at such applications, to protect the pump components by measures, such as hard overcoatings.

1.1.2 Overload

The wrong lay-out of the pump is a frequent failure cause. A higher solid matter quota than assumed, or another viscosity of the medium - oily instead of water-like - lead to overload.

Changes at the system, like pipe diameter, additional manifolds, fittings and valves, which do not draw an adjustment of the pump after itself, can also cause problems.

The maximum negative suction head has to be observed at self-priming pumps to avoid cavitation. For operation with frequency voltage system converters the permitted limits have to be taken into account. A missing non-return valve for a timed operation with short running cycles of the pump, leads also to damages on the pressure side.

The sudden change of the direction of rotation, leads to the destruction of the impeller seat, key groove or parallel key.

1.1.3 Pumping of Fluids with Solids

Solid matters are very often pumped out in fluids. In many media at given processes, considerable solid matter are contained in the fluids.

These are in cooling emulsions in tool machinery, grinding mud (sludge), in the area of sewages or in the production of textiles.

The following parameters have a very high influence on the temporal and the quantitative incidence of the wear:

- Rotation speed of the pumps [rpm]
- Solid matter quota in kg/l or kg/m^3 [%]
- Indicator for abrasivity
- Flow [m^3/h]
- Delivery pressure [bar]
- Hardness of the solid particles [HV]
- Temperature stability
- Layer thickness [in μm]
- Operating or Process temperature [°C]

The requirements on a pump which result from the media to be pumped can exemplarily be defined as follows:

Medium:	abrasive media, water, oil, emulsion, cold suds up to pH 10, solvent
Maximal temperature:	120 °C
Minimal temperature:	5 °C
Type of the solid matters:	Dirt, chips, cast and grinding particle, sand, quartz, glass abrasion

A definition as exact as possible, of medium and operating state helps to adapt the pump or the process accordingly.

1.1.4   Pumping of Fluids with Hard Solids

Since soft solid matters, are to be pumped quite good in the medium by free flow pumps, hard or very hard solid matters can cause problems, even cause a total breakdown.

Depending on the type and concentration of the solid matters in the medium to be pumped, it must be decided, which measure the best solution provides. There is no ideal solution. Depending on application {employment} different measures like rubber coating, surface remuneration, inlets of ceramic, plastic overcoatings or hard cast can offer the solution.

While rough particles are leading to signs of wear out mainly to impeller and volute, do grinding particles, sand or glass abrasion cause damages to bearings, shafts and mechanical seals.

Pipe diameters and flow speed have a very big influence on the degree of the damage and the length of the service life of the pump. The bigger the pipe diameter and the less the flow speed (velocity), the less is the damage and the higher is the service life.

1.1.5   Faulty Running and Operating

An operation which does not match the operating point necessary for the application, can have very different reasons. A wrong lay out can considerably lead to the running of the pump in the extreme case outside the operating point. Harmful vibrations with damage of bearings as a consequence, or overload and destruction of the motor can be the result. If agressive media are used, the wrong material can cause corrosion and leakages. If submersible pumps in the "slurping operation" are pumping out of a container, too much air can lead to dry run and failure of the mechanical seal.

A wrong piping can increase the decrease in pressure unnecessarily and also cause overload. If the solid matter quota is too high, the pump can be blocked, failure in pump power output or a total fail. If it comes to defects with the pump, then it is very often not a problem with the pump, but a system error.

1.2   Wear Out through Abrasion

An abrasive wear out is defined as a kind of micro shaving at which it comes to the abrasion of material on components. The hard, abrasive material in the pumping medium is harder than the material of the pump components.

Abrasive media are in lot of liquids which are delivered through pumps. Depending on solid matter the service life of the pumps gets considerably reduced, partial by harms to impellers and volutes, sometimes within only a few weeks. Under the aspects of Total Cost of Ownership (TCO) and Life Cycle Costs (LCC), this means a drastic rise of the operating costs.

In the operation of tool machines, the centrifugal pumps are very frequently used for the transportation of cooling emulsion loaded with solids, abrasive liquids. Metal shaving, grinding dust and corundum rub at volutes and impeller so strongly, that the pump fails after a short period of time.

Examples from the Practice

In the textile chemistry on textile cleaning abrasive solid matters are found in the sewage, for example at the refining of jeans.

In the civil engineering for example the abrasive lubricant "bentonite" is used. Bentonite is a rock mixture of different clay materials (among others: quartz, mica, feldspar etc.) and is applied for sealing up buildings and in the field of dyke building, and as lubricant for tunneling construction. The abrasive effect on pumping out of liquids with bentonite pollutions is considerable (see figure 3).

Figure 3 Centrifugal pump installed to a system for pumping bentonite

In the field of solar cell manufacturing (photovoltaics), submersible pumps are used for extracting the grinding muds (Slurry) when sawing silicon wafers. These grinding sludges consist of a mixture of glycol and silicon carbide as abrasive compounds (see figure 4). The wafers (thin silicon wafers) are sawed with a steel wire from the casted silicon blocs. The Operation is a permanent circulation of the abrasive compound "silicon-carbide/glycol" in the 24 h, when pumping the medium. The abrasives quota amounts to approx. 20 weight %. The permanent circulation is necessary, otherwise the abrasive compound gets hard.

Continued

Examples from the Practice—cont'd

Figure 4 Spiral volute after test with grinding muds "Slurry"

At all these applications the service life of the pumps is reduced by this abrasive wear out considerably.

1.2.1 Impeller

The material erosion on the impeller takes place at very different spots. Flow simulations pointed out very clearly, that the damages are strongly related to the velocity of the fluid (see figure 5). The greatest abrasion takes place on the impeller surface, where the fluid has the highest speed. Furthermore, an increased erosion appears on sites of a turbulent flow of the medium (see figure 6).

Figure 5 Damaged impeller by abrasion

Should abrasive media be transported, it is recommended to use pumps with an open impeller. Using free flow pumps for pumping solid matters with a particle size of 3-5 mm, with a maximum solid matter quota of 5-8 %, it is possible to prevent wear out on impeller and volute of the pump. At higher solid matter quota, this can lead to blockage or get the impeller vanes grinded down completely.

Figure 6 Impeller before and after the damage – material erosion by abrasion

1.2.2 Volute

The most material erosion takes place on edges inside the volute. For example: at the emptying hole or the drill on the wear out plate. In addition, even a similar damage takes place inside the volute with fluids containing grinding particles.

A phenomenon which can appear is an abrasive surface, arising out of the flow, according to the principle of flow optimized shark skin as shown in figure 7. The assumption that a surface as slippery as possible has the lowest resistance is wrong. By the lengthways grooves on the shack the crossways flow is lowered and therefore reduces the friction. This principle is already applied successfully in the aircraft industry.

Figure 7 "Shark skin surface" in the pump volute

If the wear out takes place so evenly, the wall thickness reduced can absolutely withstand, so that the service life of the pump can be acceptable. Concerning the flow effect of this surface "shark skin", it is even a loss minimization since this surface is actually ideal. If the wall thickness is appropriately high enough, this type of wear can shorten the life cycle of the pump only insignificantly.

1.2.3 Bearings

An overstress of the bearings occurs, if the radial force is too high. If the pump is driven with a considerably higher discharge than scheduled, it comes to the flexing of the shaft with a following damage done to the bearing. Also by strongly abrasive media the bearings of pump and motor are hurt as documented in following illustrations as shown in figure 8. Grinding mud seems primarily very strong to edges and splits.

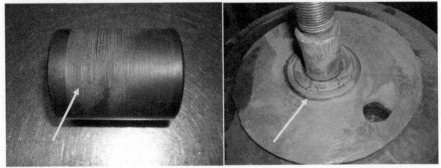

Figure 8 and 9 Damaging of the throttling bush - consequence: damage to the motor-bearing

Abrasive wear out also leads to antifriction bearings primarily into connection with corrosion to destruction. As a result it comes to amplified mixed friction and material erosion. The noise level increases continuously, the wear out further gains in cage simultaneously and rolling body increases up to the end, until the final destruction of the bearing as shown in figure 9.

Possible protective measures can be: protection ring for the motor-bearing, erection of a wear out sensor with supervisory system or a motor circuit breaker. A higher-quality motor as applied in explosion protection area could also be useful.

1.2.4 Piping

With sufficiently big pipe diameters, both flow losses and damages by abrasion can be reduced. If the pipe diameters are chosen too narrow for cost reasons, there is no stabilized laminar flow, but a turbulent flow with irregular crossways flows in the pipe. If solid particles are in the flowing medium, these have an effect to the inside wall of the pipe as in the case of a process of grinding. The strongest material erosion takes place on pipe narrowings, manifolds or also T-fittings. The solid matter particles bump against the appropriate places with increased force and wash or rebore the pipe more or less until finally the pipe wall breaks through as shown in figures 10, 11 and 12. This can happen very fast at thin-wall pipes. Regarding protective measures, it will be dealt on in chapter 4.

Altogether, with small tube diameters investment costs could be reduced, but energy expenses cannot be saved.

The increased pipe frictional resistance in pipe wit smaller diameters compared to pipes with a wider diameter, still strengthens by a high roughness in the tube, brings about a rise of the decrease in pressure. This means more pumping capacity and more energy consumption is needed. To the gentle operation of a system, for the reduction of wear out and also for the minimization of the noise level, one measure definitely is the enlargement of the pipe diameters.

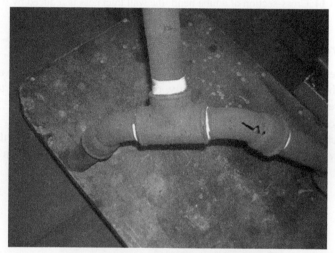

Figure 10 Frayed tube bends

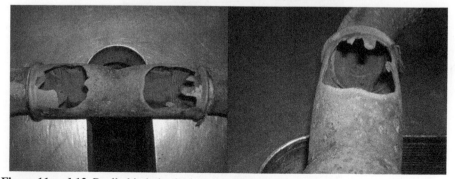

Figure 11 and 12 Really big holes in the tubes (abrasion)

On abrasive media, it has to be paid special attention therefore on the strength and resistance features. Relevant harms have to be expected also on fittings like valves, slides, measuring sensors if the materials are not sophisticated enough (see figure 13 and 14).

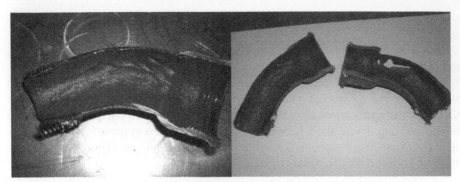

Figure 13 and 14 Cross-section of a tube bend: the wall thickness is reduced by wear drastically

1.2.5 *Abrasion and Corrosion*

If solid matter particles are transported in a corrosive medium, it comes to an overlapping from abrasion and corrosion. Usually, the abrasion rate is not easy to forecast. This is, however, very high on too high flow velocity and turbulent flow.

The service life of a pump can vary significantly, depending on grinding medium at grinding machines, for example. Depending on composition, granulation of the grinding sand and flow speed, the service life can be increased by 20-40 % by reduction of the flow.

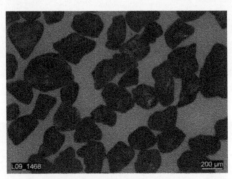

Figure 15 Grinding grains of sand, new

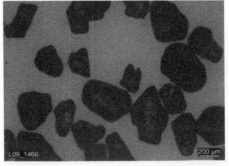

Figure 16 Center punch, sharpened

The figures 15 and 16 show the center punches of the grinding sand used at grinding processes. The left figure (15) shows the sand newly, in the right figure (16) the edges and corners of the grains of sand are already "rounded off" after approx. 100 hours bows. Depending on how often the grinding sand is exchanged in the process, the service life of the pump can therefore varying significantly.

Figure 17 Wear out by abrasion and corrosion at the impeller

Once deepening in the surface have arisen due to material erosion, the further destruction can take place very fast. Due to deepening, a turbulent flow arises which still accelerates the material erosion.

From the surface, damaged by corrosion, the material pieces breaks out of a superimposed abrasion and are washed away by the medium. By renewed corrosion and following abrasion, the material is washed out until the break or formation of holes as shown in figure 17.

The amount of the pH-indicator of the fluid medium is very decisive for the damages on a superimposed corrosion. If the pH is very low (< 5), the passive layer on the surface is destroyed and it comes to pitting.

Intercrystalline corrosion can arise when acids are pumped. The acids destroy the metal structure of the material and let the material corrod from inside out.

A fluid medium is uncritical with a pH of 6.5 - 7.8. The so-called Pourbaix diagrams which paraphrases a fundamental permanency in different pH areas also give a statement about this.

The material abrasion on critical sections in the spiral volute, such as at the mating surface spiral volute/lid, leads to leakage as shown in figure 18 and 19. In principle can be said, that the risk of the corrosion at high-grade steel/-cast also on salt-/ seawater increases as follows:

The high-grade steels are conditionally noncorrodible [14]:

Figure 18 Wear out by abrasion and corrosion at the volute

- 1.4003, 1.4016 no chloride and sulfur dioxide strain
  Resistance class 1, ferrite structure
- 1.4301, 1.4541 without considerable chloride and sulfur dioxide strain
  Resistance class 2, austenit structure
- 1.4401, 1.4571 moderate chloride and sulfur dioxide strain
  Resistance class 3, austenit structure
- 1.4539, 1.4462 (cast steel) high corrosion load (seawater/saltwater)
  Resistance class 4, ferrite-austenit structure (1.4462), austenit structure (1.4539)

Figure 19 Corrosion in the spiral volute

For the exact determination of the resistance of components a salt spraying test has to be carried out. This is an artificial ageing test or a "rotting test" at which a faster corrosion load is feigned. However, the corrosivity or the composition of the medium to be pumped should be known exactly.

To investigate the corrosion performance of metallic materials more precisely, chemical corrosion experiments and salt spraying tests could be carried through, according to the Norm DIN 50021 or DIN EN ISO 9227. Therefore, relevant influence parameters, which affect the corrosion performance can be determined with that.

1.3 Wear Out by Cavitation

Usually, cavitation appears together with a very strong noise level and with vibrations. The vibration leads to damages on shaft and bearings. Since cavitation means a strong overload of the system, it can lead very fast to the breakdown of the pump but at long last also the motor. Impellers, which have been damaged by cavitation are mostly useless, however, they can be repaired, depending on the degree of damage.

Figure 20 Cavitation in the spiral volute

The strong decrease in pressure in the flow, accompanying the cavitation, leads to evaporation of the fluid medium. The vapor locks collapsin again produce pressure blows and noise.

In the spiral volute, the particles partly break away on edges as shown in figure 20.

Overload and frequent operation of the pump outside the operating point leads to vibrations, missing strain on shaft and bearings and from which resulting to damages. Impeller and spiral volute can be completely destroyed by the strikes which results from imploding the vapor locks as shown in figure 21. The imploding vapor locks act as bullets due to the strong difference in pressure.

On the causes and emergence of the cavitation will be described in chapter 3.2.

Figure 21 Cavitation on the impeller

1.4   Wear Out on Mechanical Seals

Wear out on mechanical seals can have multiple reasons. In comparison with other components of the pump, the mechanical seals are components which have one of the shortest life time, i.e. these must be renewed again and again.

Dry up due to lacking lubrication especially on hart matings of the gliding surfaces, leads almost immediately to destruction. At soft/hart matings is the behavior insignificantly better, however, brings a shortening of the life time. Most wear out occurs on dry run, followed by a normal wear out, material wear and wrong installation.

Considerably more than 10% of the pump disturbances can be explained by dry run.

Double seals can reduce the risk of harms, since by the failure of the first seal, the second one can still resist.

1.4.1   Categories of Materials

Mechanical seals are put together out of differently components which are manufactured from very different materials. Seal ring and mating ring are produced from synthetic carbons, metals, metal oxides, synthetic materials, and carbides as shown in figure 23 and 24.

Bellows and O-ring are manufactured from an elastomer, the spring and the bracket ring are made of high-grade steel.

The basic construction of a mechanical seal is illustrated in the following figure.

The double seal type "Tandem" (Fig. 22) is used at difficult application cases of the process industry. For example if a leakage would cause very big damage. With the use of "Tandem", leakages are drained back to the system.

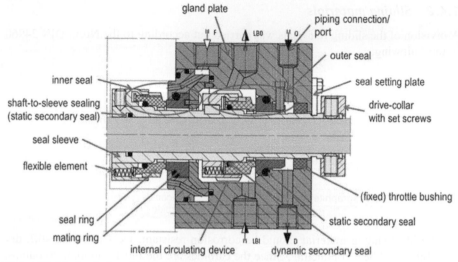

Figure 22 Mechanical seal type "Tandem" [43]

Figure 23 Bellows with feather [1]

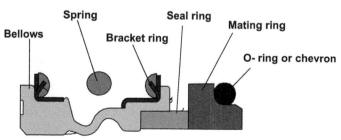

Figure 24 Construction of the mechanical seals

1.4.2 Sliding materials

A division of the sliding materials was carried out according to the Norm DIN 24960 in the following:

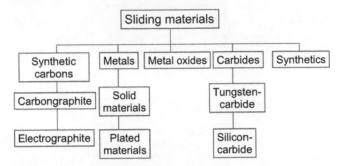

The ideal sliding material should be corrosion-resistant, wear resistant, stiff, dry run ability and heat conducting. Since the carbides are the most important to pumps, silicon carbide (SiC) and tungsten carbide (TC)/wolfram carbide (WC) shall be surveyed here a little bit more detailed here as shown in table 1 and 2.

The qualities of the tungsten carbides can be summarized as follows:

- high tensile strength
- hard, wear resistant
- low brittle fracture risk
- overheating leads to radial to warmth stress cracks in the gliding surface
- selective corrosion of the binder quota, principal in media with low pH-indicator leads to structure relaxations and increased leakage
- little dry run ability

Table 1 Comparison of the Gliding Materials Silicon Carbide and Tungsten Carbide

Characteristics	SiC	TC (WC)
Corrosion resistance	++	-
Wear resistant	++	++
Strength of shape	++	++
Failsafe running functions	-	-
Heat conductivity	++	+
Price	++	+

SiC is better in some characteristics than TC but it is also more expensive.

Table 2 Hard/Soft Classification of the Gliding Materials

Hard Sliding Materials	Soft Sliding Materials
Metals	Carbon graphites
Carbides	Synthetics
Metal-oxides	

The characteristics of the silicon carbides are defined as follows:

- very good chemical constancy at pure SiC
- hard, wear resistant
- high heat conductivity
- dry run ability can be improved by the use of carbon containing composite materials
- -overheating leads to netlike Warmth stress cracks in the gliding area surface
- selective corrosion of the free silicon, principal in media with low pH-indicator leads to structure relaxations and increased leakage
- high brittle fracture risk at shock stress

The selection of the gliding matching of materials has to be carried out depending on application and strain. Their criteria will be described in the following chapter.

1.4.3   Comparison of the Material Characteristics

The materials of the mechanical seals have very different characteristics. In the following they are compared with regard to heat conductivity, hardness and corrosion resistance (see diagram 1 and 2):

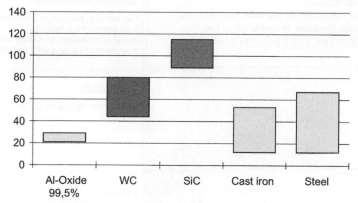

Diagram 1 Heat conductivity [W/mK] - a comparison

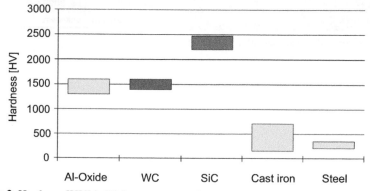

Diagram 2 Hardness [HV] in Vickers - a comparison

As obvious from the previous graphics and diagrams, SiC is the most outmost material in its characteristics, which are most important to pumps. SiC is therefore very often used in processes where wear out appears.

New designs aim at a higher dry run strength. Pure, micro crystalline diamond coatings are evaporated with a layer thickness of 6-8 μm on the SiC-gliding surface.

Therefore it is reachable, that the coefficient of friction is 5 to 8 times lower than at SiC and the heat generation consequently is rigorous reduced.

The layer survives through dry run phases which can be as long as up to an hour. Since these high-quality mechanical seals are even much more expensive than the conventional ones, they have not yet spread so strongly yet. If one, however considers the costs for loss of production, the use of these seals can absolutely be economical.

In the following, several different cases of damage are described with solution references (see figure 25 and 26).

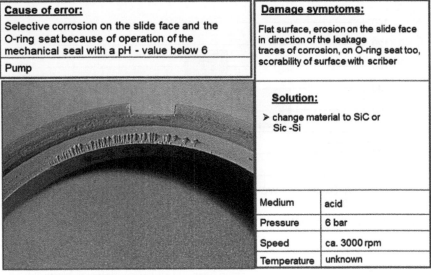

Cause of error:	Damage symptoms:	
Selective corrosion on the slide face and the O-ring seat because of operation of the mechanical seal with a pH - value below 6	Flat surface, erosion on the slide face in direction of the leakage traces of corrosion, on O-ring seat too, scorability of surface with scriber	
Pump		
	Solution:	
	➤ change material to SiC or Sic -Si	
	Medium	acid
	Pressure	6 bar
	Speed	ca. 3000 rpm
	Temperature	unknown

Figure 25 Case of damage "selective corrosion"

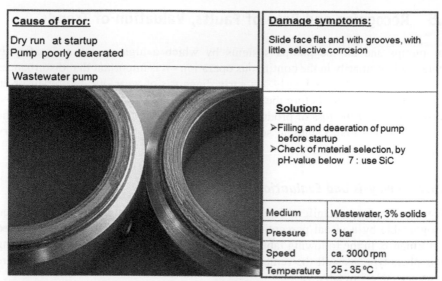

Cause of error:		Damage symptoms:	
Dry run at startup Pump poorly deaerated		Slide face flat and with grooves, with little selective corrosion	
Wastewater pump			
		Solution:	
		➢ Filling and deaeration of pump before startup ➢ Check of material selection, by pH-value below 7 : use SiC	
		Medium	Wastewater, 3% solids
		Pressure	3 bar
		Speed	ca. 3000 rpm
		Temperature	25 - 35 ºC

Figure 26 Case of damage "dry run"

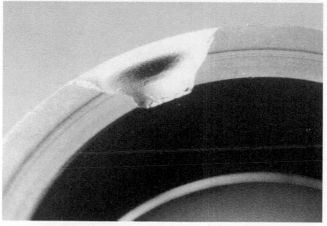

Figure 27 Damage at the installation: Warmth stress cracks in SiC

In summary, it can be said, that the most frequent disturbances which arise are dry run, overheating and warmth tensions as shown in figure 27. Thereafter arise stress cracks which draw afterwards a destruction of the sealing. If the Mechanical Seal gets in contact with abrasive particles, this leads also to the destruction. To ensure, that the leak tightness is preferably long, it is recommended, that the boundaries of application have to be maintained. This includes sliding speed, sliding pressure, temperature and friction coefficient, which substantially determine the friction at the respective seal.

1.5   Recognize and Rate of Faults, Valuation of Trends

For pumps and all the pumps systems by which a high operational safety is demanded - primarily in the continuous operation - it is important that disturbances are recognized and solved early. An early diagnosis of wear out, the repair or an exchange of the pumps before the damage, reduces high expenses, which might be caused in case of the loss of production. It is not always very easy to diagnose the exact cause and find a suitable solution. Therefore a systematic analysis is very helpful.

1.5.1   Analysis and Evaluation of Causes

Wear out on pumps manifests itself very differently. These signs of wear are often recognizable by a visual perception, noises, restless run, vibrations or decreased performance. In the following tables, the different disturbances are described using their phenotype, as well the remedying possible causes and recommendations are given see table 3 to table 7.

Very often, however, result for disturbances, which finally lead to wear, arise not in the pump themselve, but due to wrong operation, defects in the systems, plants or in the complete system.

Table 3 Identification of Disturbances by Cause Analysis

Problems	Cause
Faulty Mechanical Seal	• Dry run at putting into operation
	• Wear
	• The sliding surfaces stick together (longer downtime)
Leak	• Shaft seal worn out
	• Pump badly aligned
Pump runs restlessly	• Bearing faulty - Cavitation
	• Wrong direction of rotation
Low discharge and low pressure	• High air amount in the delivery medium
	• Motor runs only on 2 phases
	• Wrong direction of rotation

Noises

Differences in the noise level suggest very different wear out appearances.

Table 4 Analysis of Noises

Disturbance	Description	Possible Cause	Remedying
Mechanical noises	Grinding, sharpen, clatter	Wear of bearings, setback valve	Changing or throttling of pump
	Marbles	Foreign elements in the housing, impeller or piping	Removing of foreign elements
Flowing noises	Guggling	Air in the system, gas formation	Bleeding
	Roaring in pump or piping	Pump performance to great, piping too small	Check of delivering data, smaller pump, larger piping
Cavitation noises	Drumming	Cavitation	Throttling pump, greater Pump, with 4 poles motor

Table 5 Resonance Noises and Leak

Disturbance	Description	Possible Cause	Remedying
Resonance noises	Buzzing, roaring swinging	Delivery too far in the overtaxing (overloading)	Check of delivering data, changing pump
		Several pumps on a console table	Change of impeller, put vibration (shock) absorber under pump
		Rigid connection of pump and system, container shape	Separate set up, flexible piping connections, Strengthening container
Leak	Trickle between pump and motor	Mechanical seal faulty, wear, dry run, wrong direction of rotation, wrong installation, chemical decomposition	Changing mechanical seal
	Trickle between flange connections	Braced installation, old seal	Flexible piping connections Changing seal

Decline in Performance

Although the pump delivers medium, the wear out is automatic if a faulty operation is not corrected. It comes surely to the failure of the pump operation, sooner or later. If the pump power output is lower than the debit value, you must go ahead according to a check list to discover the cause. Such a possible check list arises from the following table.

Table 6 Analysis of the Decline in Performance

Disturbance	Description	Possible cause	Remedying
Pump performance too weak	Wrong direction of rotation	Wrong electrical connection	Two phases swap (exchange)
	Pump or piping stops up	Suctioning and deposit of dirt	Cleaning pump or piping, installing suction filters
	Air in the system	Air aspiration by a too low covering	Increasing fluid level, changing suction pipeline, installing level switch
	Setback valve (flap) does not open	Air in the system	Vent pipeline directly in front of setback flap
	High air fraction in the medium	Process of machining	Degassing of medium

Fault Analysis by Means of Electrical Data

Besides mechanical disturbances and interferences, of course electrical damages are also a result of faulty operation.

Table 7 Analysis of the Electrical Data

Disturbance	Description	Possible Cause	Remedying
Overload of the motor	Electrical consumption too high	System resistances too low, faultily manometer, electrical connection wrong	Throttle pump, smaller pump, smaller impeller, compare electrical connection with motor type plate, Use larger motor
Tension at the clipboard	Blockage	foreign elements, deposits	Cleaning
	Rotatable	Winding faulty	Change of motor or stator

1.5.2   Evaluation of Trends

If at the beginning into operation, there are no disturbances and no faulty operation conditions to be recorded, it can almost be excluded, that there is a fault or error in the system or in the pump.

In the course of the operation of a system the operating state can change, however.

A modified medium, other operation, a changed characteristic curve of the plant by reorganization measures in the system, can lead to disturbances and consequently to wear out and damages.

A rise of the noise level, power loss or a change of the electrical data indicates an overload or missing strain in the system.

An early diagnosis of trouble, by using prognostic maintenance systems is reasonable at complex systems in every case, since the resultant damages generally are not insignificant by loss of production.

Repairs and maintenance measures can therefore be included in one's plans well, punctual before the claim occurs.

1.5.3   Measures for Prevention of Failures

By long-standing experiences with damages on pumps it is possible to meet inferences on the causes of the damage after an exact analysis of the harm has been made. Learning from the damage images can already lead to an damage avoidance by a foresighted maintenance prior to the harm initiating itself. From the damage symptoms, measures can be derived, which minimize the wear out by prevention or specific monitoring.

Are there arising technical expertise, which has a recurring damage forecast, due to the monitoring of the operation of the pumps, constructive modifications are the most reasonable measures.

If wear out-promoting operating states, like delivering fluids with solid matters, such as cooling emulsion which is polluted with metal shaving or grinding dust, the expected damage can be predetermined with the help of a flow and wear out simulation.

Provided that constructive measures are not practicable for different reasons, other possibilities of the product adaption must be found.

Surface coatings, which in this respect change (mutate) the surface layer of the components to become harder, i.e. more resistant to solid matters, offer a refining of the components as an option.

1.5.4   Flow Simulation

To be able to analyze the flow of the medium which is delivered, in the impeller and spiral casings more exactly, the flow course can be calculated by using corresponding software (CFD, solid works, and others) as shown in figure 28. Therefore, exact geometric data to the pump as well as the flow data like flow rate, pressure etc. must be available.

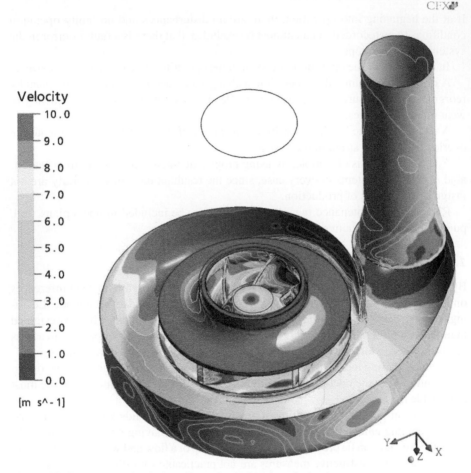

Figure 28 Speed spreading in impeller and spiral casings (volute)

For the calculation, the geometry data of the pump are processed with the software. A networking is then carried out as a preparation for the computation of the flow.

By the simulation of different operating states a damage image can be made.

Through tests it has been shown, that the damage images of the individual components corresponds very well with the results of the flow simulation. The areas with an increased flow velocity show the strongest signs of wear out.

Based on the results of simulation, constructive measures or component changes (alterations) can be derived. In chapter 3.5. "flow optimization" it is described more exactly, how the simulation of wear out due to abrasive fluids is possible, by means of flow simulation.

1.6 Damage Images

Due to the damage images, inferences on the causes of loss do not always tell the reasons. An analysis of all factors of influence must be systematically carried out to be able to exclude possible causes. A flow and wear simulation can be very helpfully by means of programs which are available on the market.

Examinations showed, that the damage images of particular components are almost congruent with the results of the flow simulation. The places with an increased flow velocity show the strongest signs of wear out. Wrong lay out of the system (cavitation, damage of bearings), too high solid matter quota (overload, blockage), too hard solid matters (abrasion + "holes") are often the cause of the damages. Deposits accumulating over a longer time period, primarily for discontinuous operation or for operation with quite a long standstill time, lead to damages. Here, for continuous operation, substances which could lead to deposits, are mostly rinsed out from the pump.

Deposits must be avoided, since they can stop pump and pipes up. In extreme cases they can also cause a chemical decomposition of the material.

Figure 29 impeller with sticky varnish (paint) residues

Figure 29 shows the impeller of a varnish (paint) coolant pump. Too low flow speed, a too high varnish amount, or a too long standstill time, caused the sticking of the impeller and thereby leading to total failure of the pump.

By repair, the harm should always be analyzed exactly and not only the faulty component replaced. Direct measures for the prevention have to be initiated.

It should be checked, whether constructive modifications of the system or the pump must be arranged. Frequently recurring damages conclude faults at operation or by design.

2 Measuring, Monitoring and Documentation of Faults and Wear Out

On monitoring of systems or pump characteristic, in principle it has to be decided, which effort for it is adequate altogether. Whether permanent, or with a mobile measuring system only from time to time, whether measuring is cyclically or daily or one carried out only weekly is sufficient. There are the fitting sensors for all process parameters like speed, temperature, pressure, difference pressure, vibrations, electric current and voltage, flow rate, leakage and filling level to be measured; up to the permanent monitoring by lasers of the concentricity of a shaft. Primarily, the measuring and the documentation of the alteration of a measurand is necessary. The most important parameters, to record the condition of a pump, will be described here more precisely. These are pressure, temperature, speed, electrical current and vibration.

2.1 Vibration Measurement

For the vibration examination, measurements are carried out by means of so-called impact sound sensors. The sensor, when executed as acceleration sensor, measures the vibration in terms of the pump in g (earth acceleration: 9.81 m/s²). The sensor is either screwed on tightly at the spiral casing of the pump, or fastened with a magnet foot, for the mobile measuring or monitoring. The results of the impact sound vibration measuring are very meaningful. The vibration value measured at the pump, can be indicated directly to the damage symptom.

2.2 Temperature Measurement

Temperatures can either be measured with thermocouples or with PT 100 -sensors. Test points are critical pump- or plants-components like mechanical seals, ball bearings, motors, pipes as well as the delivery medium on the pressure or suction end.

A noticeable increase or change of the temperature indicates an error or a wear out initiating itself. A temperature monitoring is already integrated in many electric motors, to the protection from overload (PTC thermistor).

By overload, the temperature in the clamp box of the motor is higher than the temperature directly at the pump (bearing, spiral casing) since the motor is heated up faster by overload. For the temperature monitoring, the electronics in the clamp box of the pump is examined in a fixed period of time (e.g. in second stroke). For example, a permitted maximum value of 100° C will be determined. If the temperature exceeds

Damages on Pumps and Systems. http://dx.doi.org/10.1016/B978-0-444-63366-8.00002-4

the predefined configuration-worth, a temperature fault will be indicated. If this fault remains on permanently, the pump should be switched off.

Depending on the specific application and the used monitoring system, the temperature is controlled and recorded in the medium or in the clamp box of the motor.

2.3 Pressure Measurement

The monitoring of the delivery pressure establishes guaranty, that the flow of the fluid remains constant and does not stall. Pressure fluctuations, pulsations, pressure surges and also negative pressure can be controlled and recorded by corresponding systems. The pressure measuring is carried out on both sides, on the suction side and on the pressure side, to get meaningful values about regarding the decrease of pressure. A differential pressure transmitter registers the difference in pressure between inlet and outlet.

2.4 Speed Measurement

Depending on the application used, the rotary speed shall remain constant or be regulated for example with a frequency converter, in order to adapt the needed pump power output. For the stationary monitoring, the speed can be readout as an analogous signal from the electronics in the motor clamp box. For short-time mobile measuring, a laser gauge is used. Therefore, a reflection mark is attached on the rotating fan impeller. This mark is red by the laser sensor at every turn of the fan impeller of the motor.

If the speed is too low, an overload can be diagnosed. If the speed is increased the reason might be that the inlet end is partly or completely closed. The speed is increased strongly, if the air quota is too high. Cavitation can be diagnosed, if the speed is absolutely very high and the pump is noisy.

2.5 Electrical Current Measurement

The electric current of the motor can be monitored by a minimum/maximum recording system. A deviation is diagnosed as a fault. In idle speed and at dry run, the pump takes less than half of the debit current in comparison with the normal operation. The monitoring of the current is configurable and can start a few seconds after the start and last until the switch off of the motor.

2.6 Damage Diagnosis by Condition Monitoring and Vibration Analysis

The real examination of the vibration measurement forms the fundament of the failure analysis. By the vibration analysis it can be concluded so exactly than in no other process, about the type of the damage or on a faulty operating. Every pump has a typical vibration behavior of its own which changes at different operating states. By vibration analysis, diagnoses can be named very precisely like bearing damage, balancing fault, wear to impeller blades or also cavitation.

By frequency analysis the damage images can be distinguished very exactly from each other. The permanent condition monitoring of the vibration makes sense, at critical processes, for very expensive pumps and systems, where no replacement pump is kept as stand by.

process about the type of the damage or an actually operating. Every pump has a typical vibration behavior of its own which changes at different operating states. By vibration analysis, diagnoses can be named very precisely like bearing damage, but ancient fault, wear in impeller blades or also cavitation.

By frequency analysis the damage images can be distinguished very exactly from each other. The permanent condition monitoring of the vibration makes sense in critical processes, for very expensive pumps and systems, where no replacement pump is kept in stand by.

3 Prevention of Cavitation and Wear Out

The most important measure for the prevention of damages is the avoidance of lay-out and planning errors. The exact knowledge of the system operation, the availability of the system scheme makes the choice of the pump very much easier.

The disregard of the constitutions of the hydraulics, often for cost reasons leads innevitably to damages in the pump and in the system. Foresighted maintenance and the use of a monitoring system should be considered. In the long run, the flow optimization of the system offers protection against wear out in any case.

3.1 Prevention of Errors in Layout and Planning

Costs can be saved mostly by precise planning, design and exact construction. Subsequent costs which arise only after some time can be avoided by suitable measures. Particularly, these are in detail:

- set the exact operating point (system characteristic curve, pump characteristic curve)
- consideration of flow losses
- check sucking requirements (suction pipeline, whether self-priming)
- adaptation of operating point (speed control, frequency inverter)
- consideration of viscosity (water, oil, other fluids)

If the parameters are defined and registered, further steps can be realized by means of different software.

3.1.1 Exact Definition of Operating Point

Computer programs for pump selection, offer choice of pump size, according to the correct determination of the application.

Delivery flow and rate, impeller diameters, type and size of motor, use of a frequency converter are defining criteria. If the system characteristic curve is not known, the characteristic has to be investigated by calculation of pressure loss of the pipeline system including all the different fittings.

The best operating point normally lies in the area of the highest efficiency factor (see diagram 3 [24]). This means an optimal setting of pressure, delivery flow and electrical performance. If the pump is driven in this operating point, a very gentle operation can be ensured.

Damages on Pumps and Systems. http://dx.doi.org/10.1016/B978-0-444-63366-8.00003-6

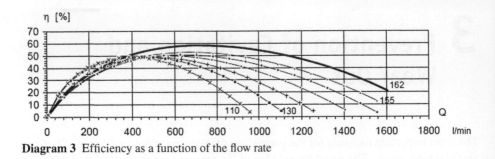

Diagram 3 Efficiency as a function of the flow rate

For the pre-selection of the pump size, the characteristic diagram (map) has to be viewed (see diagram 4 [24]). It should be avoided to operate the pump at the range limit of its pump operating map. Often, the information from the customers or the operators of a plant are often not quite exact. It is recommended therefore, to provide power reserves.

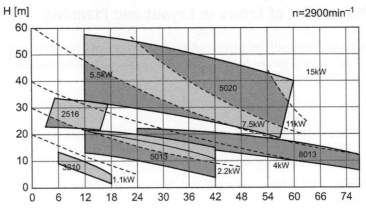

Diagram 4 Pump operation map: Total head as a function of the flow rate Q [m³/h]

The Q-H-characteristic is one of the most important selection criteria for the choice and determination of the pump size. The necessary total head H and the desired volume or flow rate Q form the base for the exact choice of the pump.

Further details on the delivery task of the pump are required in addition. Data to pH-indicator, viscosity, corrosivity temperature stability are urgently also necessary to prevent damages. If the pump must be self-priming, this is also an important criterion.

3.1.1.1 Operating Point and System Characteristic Curve

The operating point adapts itself to the intersection point of these two lines in dependence of pump characteristic curve and system characteristic curve automatically (see diagram 5 [24]). For the choice of the right pump, the system characteristic curve or data must be known about it.

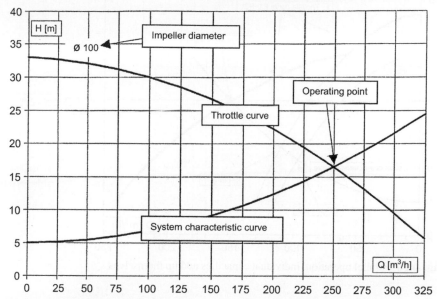

Diagram 5 Pump characteristic curve (throttle c.) and system characteristic curve

The system characteristic curve arises from a static and a dynamic part.

- Static part: H_{stat}
- Dynamic part: H_{dyn}

The static part consists of the geodesic height and the pressure drop of the system.

The dynamic part takes into account the flow losses between inlet and outlet of the system, which rise with an increasing discharge. For the minimization of the flow losses, a flow velocity of $v = 2$ to 3 m/s is recommended. Correspondingly, the pipe diameters have to be selected.

If a desired capacity shall be achieved with several pumps, the pumps can run in the parallel operation (see diagram 6 [24]). The capacities add themselves up, however, the operating point adapting, must be checked.

The diagrams in the followings show operation in series and parallel operation.

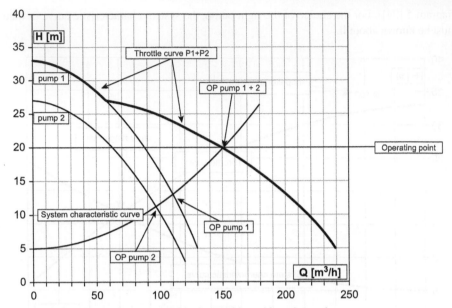

Diagram 6 Parallel operation, individual capacities add up themselves

If the delivery pressure shall be increased, pumps can be connected in series (see diagram 7 [24]). The single capacities do not add up themselves up (flow throttling on operating point P1+ P2 normal level).

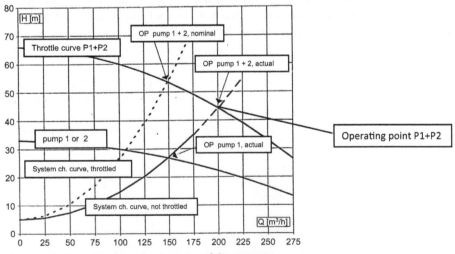

Diagram 7 Operation in series, increase of the pressure

If the pumps are not tuned with each other optimally at the two modes, parallel and series operation, flow losses are generated and in extreme cases it comes to damages.

Different software programs for the pump choice are available on the market.
In the following a short description:

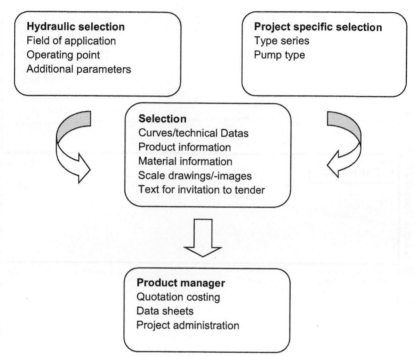

Figure 30 Example of a program for pump-selection

It is prerequisite for the exact pump selection, that the range of application, the system characteristic curve, the exact operational parameters as well as the expected pump type are known. For example self-priming, submersible pump, delivering medium with solid matters, might be the requirements. Many parameters as much as possible should be known.

With the help of the selection programs, datas for the suitable pump can then be found fast. This is shown in figure 30.

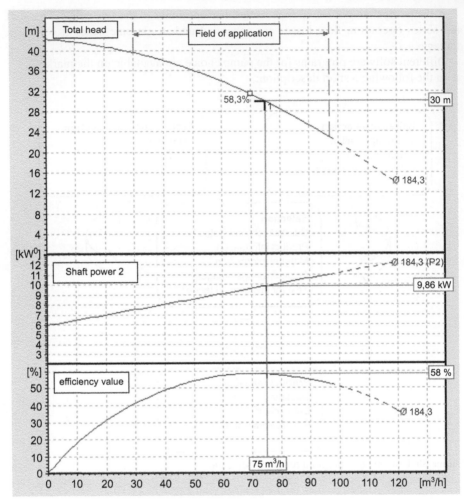

Figure 31 Results, investigated by the pump selection program

Diagrams in figure 31 [24] shows the results of a computation example, by using the operation parameters for the calculation. The desired total head is 30 m(3.0 bar), at a flow rate of 75 m³/h. The optimization is achieved by the trimming of the impeller diameter. The impeller size is reduced to a diameter of 184.3 mm. Here, the efficiency factor in the operating point amounts to 58 % at a motor power consumption of 9.86 kW. Therefore, for this operating point, the best possible efficiency factor at a optimized energy consumption can be achieved. The trimming of the impeller diameter is still more precisely described in chapter 3.1.3.2.

3.1.2 Consideration of Flow Losses

Every valve, every gate valve, or pipe manifold, even built-in sensors cost flow energy, which must indeed be taken into account at modifications to systems. The flow chart or the system scheme forms the basis for the function of the system. Larger flow losses can already be detected in the system scheme. The flow conditions are strongly affected by the choice of the pipes (internal roughness) kind of fittings, valves or slides, manifolds, elbows and compensators etc. As a rule, less dissipating, high-quality components are usually more expensive than standard components. However, a single investment in "less loss" pays off, since through this decision energy and therefore expenses can be saved (see also chapter 3.5).

3.1.3 Operating Point Alignment

To adjust the pump operation on the optimal operating point, different possibilities are available. The throttling by means of slide or valve is very often utilisized, it is however, one of the worst options. By fluidic issues, it is very unfavorable, it destroys energy and therefore costs money. Adaptation by speed control or the trimming of the impeller diameter are better and more economical measures.

Another operating facility offers the bypass regulation. Here, a part of the discharge is diverted and led back into the container. Through a control valve in this second pipe, the flow rate is regulated in the main pipe.

3.1.3.1 Variable Speed Drive by Frequency Inverter (VSD)

A very effective measure, also to the costs and energy savings, offers primarily variable speed drive by frequency inverter of the pumps. However, why does the frequency inverter contributes to the avoidance of damages? The speed control with frequency inverters is a very gentle and energy-saving measure. The exact adjustment on the optimal operating point supports a smooth run of the pump and a constant flow of the delivering medium in the pipeline system (see diagram 8 [24]). Vibrations in the system can be avoided mostly, and bearings will be conserved. Through slow start, the motor is conserved, so that it does not come to overload problems (ramp circuit).

Without frequency converters, the adjusting of the operating point is only possible directly for the pump characteristic curve, not any arbitrary point is single adjustable. Frequency converters make a direct variable speed control possible in a defined field. For the lay-out of a pump on 50 Hz this means:

Frequency f:

$f_{min} = 10$ Hz
$f_{max} = 50$ Hz

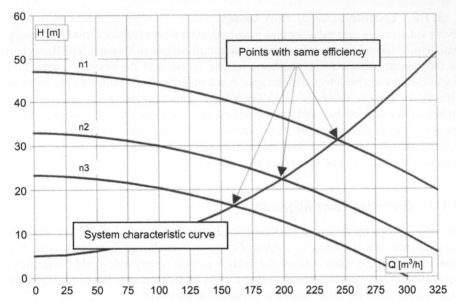

Diagram 8 Adaptation of operating point by speed drive 1

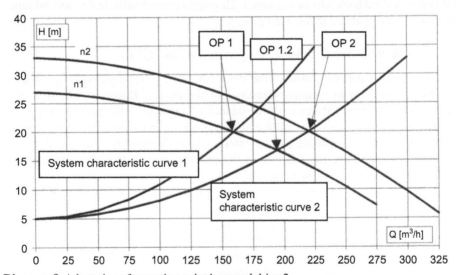

Diagram 9 Adaptation of operating point by speed drive 2

The operating point of the pump is exactly adjustable, by modification of the system characteristic curve and simultaneous shifting of the pump characteristic curve (see diagram 9 [24]).

The tuning of the flow rate by a speed control system has broader advantages. By use of a frequency converter, which adapts the speed of demand, the electricity consumption and thus also the costs proportional flow rate reduction can be saved. By suitable control, savings are possible to more than 50 %.

By rise of the frequency (e.g. from 50 Hz on 60 Hz) the following delivering data change:

- Speed (3000 rpm > 3600 rpm [x 1.2])
- Capacity ($Q_2 = Q_1$ x 1.2)
- Total head ($H_2 = H_1$ x 1.2^2)
- delivery rate ($P_2 = P_1$ x 1.2^3)

Table 8 Modification of the Delivery Data by Frequency Rise

	Capacity	Total Head	Rate/Power
n1 = 3000 1/min	Q1 = 200 m³/h	H1 = 22,2 m	17,8 kW
n2 = 3600 1/min	Q2 = 240 m³/h	H2 = 32,0 m	30,75 kW!
Factor	1,2	1,44 [= $1,2^2$]	1,73 [= $1,2^3$]

From above tables it is obvious, that a rise of frequency by 20 % up to 60 Hz, the pump power output increases by 73% (see table 8 [24]). Consequently, due to reduction of speed, the energy consumption can also be reduced drastically.

Optimal Pressure or Flow Rate at Cascade Circuit
Speed Controlled Pumps Provide a Constant Pressure or a Constant Flow Rate. For the operation with several pumps into cascade the pressure sensor records for example the pressure in the pipe and regulates the predefined nominal value depending on function (see figure 32 [24]). The signal floats directly to the frequency converter which adapts the flow rate to the predefined value.

Figure 32 Cascade circuit of pumps

For the operation with several pumps, a sensor records every nominal deviation which arise by switching-on or -off further pumps. The signal of the sensor is transferred directly to the pump and regulates the speed according to the requirement up or below (see figure 33 [24]). By the optimal setting of pressure or flow rate, the process reliability is ensured, so that the system is always driven with the desired nominal point (see diagram 10 and 11 [24]).

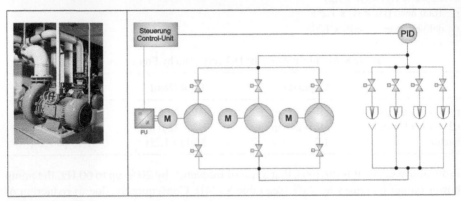

Figure 33 Cascade circuit of several pumps

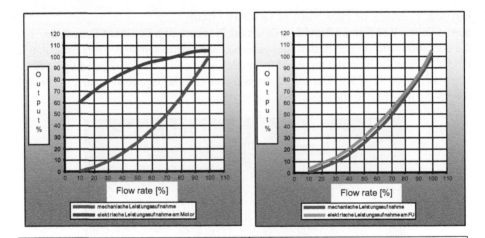

Energy requirement (output) without frequency converters At 50 % throttling (red curve), still 90% electrical power consumption (violet curve)	Energy requirement (output) with frequency converter At 50 % throttling (red curve) only 30% electrical power consumption (green curve)

Diagram 10 and 11 Power consumption for operation with and without frequency converters

The advantages and boundary conditions of the pump operation are summarized here once again:

- operating point of the pump is exactly adjustable by modification of the system characteristic curve and simultaneous shifting of the pump characteristic curve (by frequency modification)
- lower energy consumption
- an optimal efficiency factor of the motor/the pump is parameterizable
- motor preserve saving (no overload problems) by soft start (ramp circuit)
- it is valid in principle: A rise of the defined maximum frequency is not recommended
- pumps which were designed for a defined maximum frequency operation (e.g. 50 Hz) should not be changed in the frequency face-up (e.g. 60 Hz) since otherwise the motor can be overloaded or the pump cavitates
- if a frequency modification scaling up is nevertheless necessary, the impeller diameter must be reduced
- also by backfitting of the frequency converter, an overload of the motor by a faulty operation of the converter is possible

A pressure control is also practicable by means of frequency converter and conserves the pump. If the pressure in a piping network is kept constant, pressure surges and also dry run can be avoided. Through programming of the nominal parameters, the setting of the desired pump operation can be exactly adjusted.

Starting out from a total costs consideration at which the energy proportion, is with 45 % the largest cost factor (see chapter 7.3) it is very worthwhile to optimize here. Particularly since there is a good chance due to assumption of increasing prices of energy, in future with exponential increase of the costs for energy is to be expected.

In future, due to legal regulations of the European Union more speed controlled drives will be used, unregulated drives will be a rarity.

It is explained in the following example computation, how different operation intervals of a pump stipulates the economic point of view (see diagram 12 [15]).

Tables 9a - 9c Specifications for the Calculation [15]

Pump Data	
Rated working point:	1 bar / 350 m^3/h
2nd operating point at rated speed:	0.01 bar / 600 m^3/h
Rated speed:	1,430 1/min
Rated efficiency:	75 %

Data Relating to System Characteristic	
Max. operating point (for frequency distribution):	1 bar / 350 m^3/h

Data for the Energy Cost Evaluation	
Energy price:	0.1 ₡/kWh
Additional inverter costs:	2,500 €
Efficiency of the drive axis:	90 %
Alternative control:	Throttle control

In this example of calculation, a pump is operated in the operating point 1 bar/350 m^3/h (see table 9a - 9c [15]). Starting out from a total currency of 7 700 hours per annum, the pump runs in different intervals (see frequency distribution into diagram 14). An annual cycle was simulated by 500 h/a, at 10% of the flow rate, 800 h/a, at 20%, 1 300 h/a, at 30% of the flow rate and further intervals (see diagram 13 [15]).

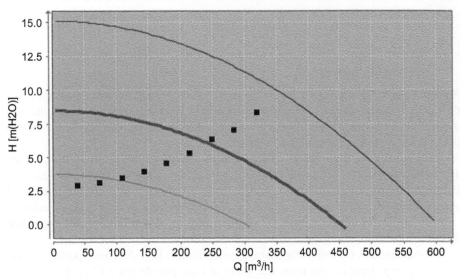

Diagram 12 Assessment operating points with speed dependent characteristics [15]

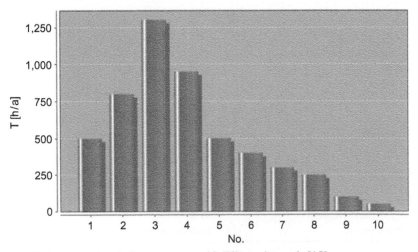

Diagram 13 Running time in h per annum at 10 different intervals [15]

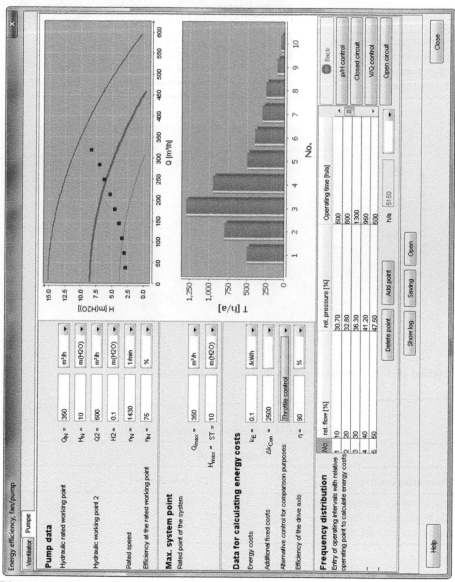

Diagram 14 Data summary for the calculation [15]

Tables 10 Result of the Energy Cost Calculation [15]

Comparison of Energy Costs			
	Annual Energy Costs	**Savings with Inverter [€]**	**Savings with Inverter [%]**
Speed-controlled (inverter)	1,506€	0€	0%
Max. working point	6,676€	5,170€	77%
Bypass control	6,622€	5,116€	77%
Throttle control	6,794€	5,288€	78%

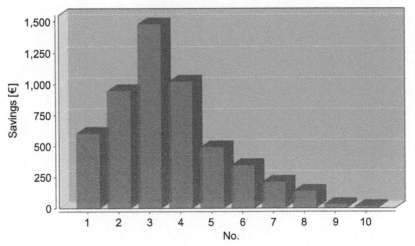

Diagram 15 Cost-cutting per annum at 10 different intervals [15]

Tables 11 Amortization Consideration [15]

Amortisation	
Alternative control	**Throttle control**
Annual energy savings	5,288 €
Additional inverter costs	2,500 €
Theoretical payback period	0.5 a

The calculation showed that the additional expenses of 2 500,- € for a frequency converter amortize themselves at this operation in 10 different intervals already after 0.5 years (see table 10 and 11 [15]). Even other much more unfavorable operation intervals, a frequency converter amortizes itself after a few years. The cost-cutting per annum is shown in diagram 15 [15].

3.1.3.2 Adaptation of Impeller Diameter

An adaptation of the pump to the operating point without additional components, offers the conditioning of the pump impeller. By the change of the impeller diameter - by trimming - the required operating point can be adjusted.

The trimming curves are described in the following diagram (see diagram 16 [24]).

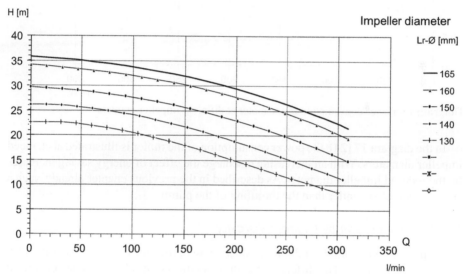

Diagram 16 Operating point adjustment by trimming of the impeller

The pump cannot be regulated directly as with a frequency converter, but with a mechanical, single measure, the operating point can be adjusted. Another later modification would mean a further trimming of the impeller or to install a new impeller.

P [kW]   Impeller diameter

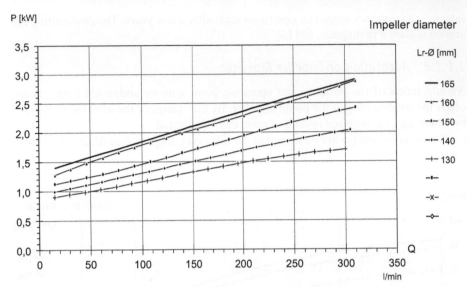

Diagram 17 Power consumption of the motor at different impeller diameters

In the diagram 17 [24], the power consumption of the motor is illustrated at changed impeller diameters. With an increasing discharge the effect of energy saving increases by the reduced impeller diameter. As described in the previous chapter already, such a adaptation is also gentler than the throttling of the pump.

3.1.3.3   Adaptation by Changed Viscosity

If a pump for oil or a fluid similar to oil shall be used, then the different viscosity must be taken into account. The different operating conditions at oil have to be considered to protect the pump from wear out or even from destruction of the pump. Operating point (OP) and system characteristic curve for oil looks different than the one for water (see diagram 18 [24]).

The higher viscosity of oil as compared with water, yields a steeper pump characteristic curve and also a steeper system characteristic curve.

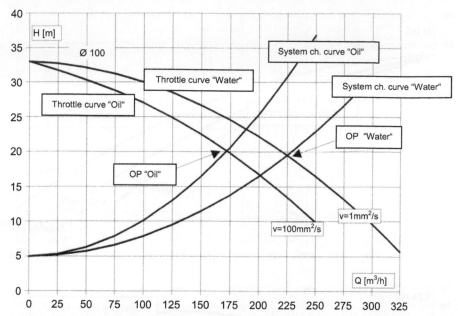

Diagram 18 Comparison of the characteristics for water and oil

The operating point shifts on the intersection point of the throttling curve for oil with the system characteristic curve for oil.

The changes as compared with water or water-like fluids can be summarized as follows:

- Viscosity: higher
- Zero delivery head: remain the same
- Efficiency: downgrades
- Power demand: increases
- Conversion: is not flatly possible

The causes for the increased power demand arise from an increased friction in the pump, the pipes and fittings.

As obvious from the diagram 19, both the total head (throttling curve) and the efficiency factor of the pump are with water higher than at oil (see diagram 19 [24]). The difference cannot be scaled down by means of a factor, but must be calculated in each individual case. Always, in the case of the same delivering conditions, another fluid such as oil, a larger pump is needed, since oil has a higher viscosity than water.

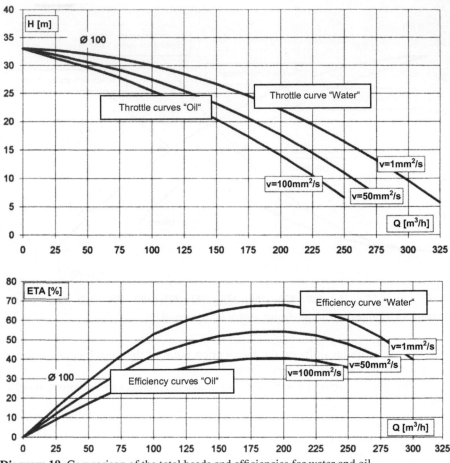

Diagram 19 Comparison of the total heads and efficiencies for water and oil

3.2   Causes for Cavitation

The causes of cavitation are local pressure lowerings, caused by over-speeds or vibrations. It can also arise from global pressure lowerings, caused by reduced atmosphere, higher suction, or reduced draw height. Friction in the system is another cause.

   Symptoms for it are increasing noises and vibration of the pump and the foundation. The impeller can be damaged at overload, at the inlet range, and at partial loads at the outlet range.

3.2.1   Harms Caused by Cavitation

If the boiling pressure is achieved or fall below in the streaming fluid, cavitation arises (Latin: cavus = cavity). By a local pressure lowering, the cavitation causes an evaporation of the delivery fluid in the system. This happens at over-speed.

Steam bubbles form themselves, which are dragged along in the streaming fluid. They condense again in areas of a higher pressure than the evaporation-/boiling pressure and collapse. The steam bubbles collapse suddenly, with sound velocity and with a high noise level, they "implode".

Pressure surge locally arise to over 1 000 bar. At this implosion, a fluid jet with high speed, a "micro jet" arises and leads to crater shaped material abrasions inside the pump or the system (see figure 34 [24]).

Figure 34 Impeller with "cavitation craters" on the impeller vane surface

Wall roughness can also lead to evaporation germs which activate cavitation. For longer operation with cavitation, wear appears also at bearings and mechanical seals.

Figure 35 New bronze impeller   **Figure 36** Damaged bronze impeller

The damaged impeller (in figure 35, the impeller is new) was destroyed by overlapping of chemical influence, abrasion and cavitation (see figure 36 [24]).

3.2.2   Types of Cavitation

The cavitation types appearing mainly into pumps are layer cavitation and cloud cavitation.

3.2.2.1   Layer Cavitation

The layer cavitation is indicated by a large coherent vapor area. Caused by a increased face velocity, a detachment of the liquid flow, takes place at the leading edge of the impeller. After a short time, this phenomenon leads to a complete detachment of the liquid flow away from the aerofoil surface of the impeller vane. The consequence is a strong decrease of the efficiency factor (see figure 37 [44]).

3.2.2.2   Cloud Cavitation

This kind of cavitation is the most frequently appearing type on centrifugal pumps which causes cavitation erosion. As soon as the appearance of the layer cavitation another rise of the flow velocity occurs, a transition enters from stationary to transient layer cavitation.

In the back area of the impeller blades, the vapor area bursts and leads to the rise of steam bubble clouds. The accumulation of very small bubbles in the size range of 10-20 μm (Micrometer) is called cloud cavitation. Again the imploding steam bubbles lead then to harmful erosion on impeller and spiral casings.

Experiment   Simulation

Figure 37 Cloud cavitation at an airfoil [44]

3.2.3   Cavitation and NPSH-Factor

The NPSH-factor of the pump paraphrases the suction head of the pump related to the center of the pump. Translated means NPSH = Net Positive Suction Head. This net suction head classifies the sucking behavior of the pump.

The required NPSH-factor of a pump specifies, how much higher the complete delivery head must be at least above the steam delivery head of the delivery fluid, in relation to the NPSH-factor, to ensure a faultless operation of the pump.

The available NPSH-factor of the system must be higher than the NPSH-factor of the pump, in every case. For safety reasons, a reserve of 0.5 m is recommended, which means that

$$\text{NPSH}_{\text{plant}} \text{ must be } \geq \text{NPSH}_{\text{pump}} + 0.5 \text{ m}$$

$\text{NPSH}_{\text{plant}}$ = net positive suction head of the plant
$\text{NPSH}_{\text{pump}}$ = net positive suction head of the pump
The NPSH-factor has the dimension m.

This NPSH of the pump is the one, which exists in the impeller inlet nozzle, and also as a required minimum inlet head necessary, so that no cavitation occurs.

The NPSH characteristic of a pump can be measured on the pump test facility (see diagram 20 [24]). The test fence works with a closed circuit. Lowering the system pressure is carried out with a vacuum pump.

The computation is carried out according to the formula:

$$\text{NPSH} = \frac{p_1 + p^{amb} - p^v}{\rho \cdot g} + \frac{v1^2}{2 \cdot g} \pm Zs$$

At which:

p_1 = *pressure in the inlet cross section of the system*
p_{amb} = *air pressure*
p_v = *evaporation pressure of the delivery fluid*
$v1$ = *Flow speed in the inlet cross section of the system*
z_s = *geodesic height, related to the reference height*
ρ = *Density of the delivery fluid*
g = *local acceleration of gravity*

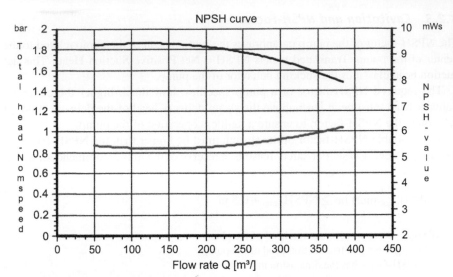

Diagram 20 NPSH characteristic measured in terms of the test bed

3.2.4 Prevention of Cavitation

Low pressures should be avoided in respect that the evaporation pressure should never fallen below pressure fluctuations. An increased suction head or a reduced inlet head must also be avoided.

The selection of material affects the risk of the cavitation erosion, therefore the possible material erosion, considerably. Therefore gray cast iron is much more sensitive than a high-quality stainless steel cast or aluminium bronze. Upcradings like surface refining, coatings or build up weldings have a very positive effect on the resistance to cavitation erosion.

Further measures help to reduce cavitation damages or to avoid them completely:

- Choosing a rotational speed of the pump as low as possible
- Sharing delivery flow on several pumps
- Installation of a pre-added impeller (see also chapter 4.5.2., Inducer)
- Keeping liquid temperature as low as possible (considering evaporation pressure)
- Choice of manifolds and fittings with small resistance coefficients
- Specific selection of material, primarily with corrosive water as fluid
- Choice of short suction pipelines with a cross-section as large as possible

3.3 Dry Run Protection

As very well known about mechanical seals, dry run due to lack of lubrication, leads immediately to the destruction of the seal. Besides systems for condition monitoring, there are systems available, at which a blocking fluid provides lubrication and therefore prevents the dry run.

3.3.1 Confining Chamber Systems

In systems without blocking pressure, the mechanical seals are charged pressurelessly with jamming liquid. The mechanical seals are installed in a jamming (confining) chamber, which is filled with jamming liquid. This liquid provides lubrication and cooling and therefore prevents the dry run (see figure 38 [24]).

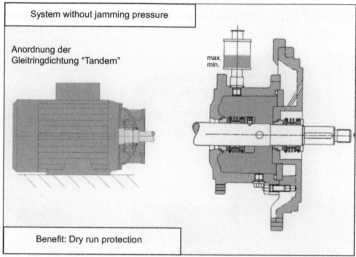

Figure 38 Dry run protection system with jamming chamber without jamming pressure

When required, the liquid in the jamming chamber is refilled. At this systems with jamming pressure, the mechanical seal is also installed in the pressurized chamber (see figure 39 [24]).

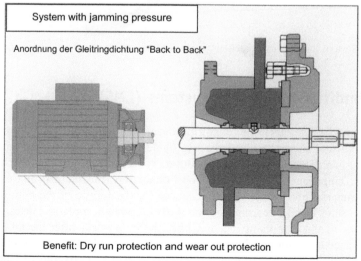

Figure 39 Dry run protection system with jamming chamber and with jamming pressure

The circulating jamming liquid is under overpressure and prevents ingression of dirt particles. This system therefore offers a dry run protection and a wear out protection.

3.3.2   Pump Control

This system offers a dry run protection by which the electric current values are supervised. As soon as at dry run the current value decreases, the pump switches off. The system still has further functions which are described more precisely in the following chapters.

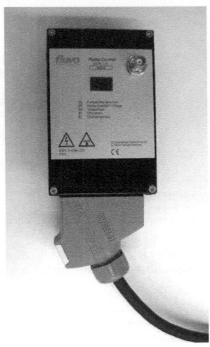

Figure 40 System for dry run protection "pump control" supervises the electric current values

3.4   Condition Monitoring Systems (CMS)

Such systems for the electronic condition monitoring make it possible to prevent the wear out and help to reduce the life cycle costs through early diagnosis of disturbance. Wear out or wrong application, causes the failure or destruction of the pumps. Unproper application, wrong connection (mechanical and electrical) or wrong material choice, are frequent causes for the failure of the pumps. Faulty mechanical seals or ball bearings are the most frequent complaint causes. Approximately 30% of the complaints occur within the first weeks after the take-off into operation. Electronic monitoring systems of pumps enable to do an exact observation of wear out and causes of damage. The proof can be delivered why and when a pump breaks down.

However, when do CMS systems make more sense? At applications for critical fluids and operating conditions: e.g. fluids with high solid matter quota, saline solutions, chemical products, critical procedure processes, or pump dry run over a longer time

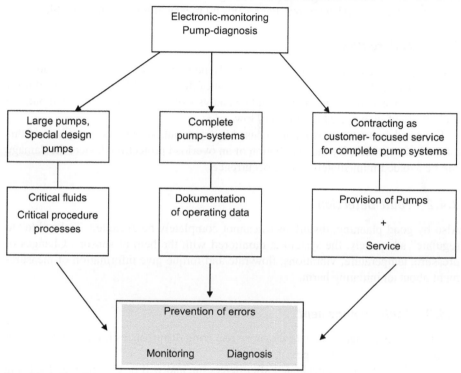

Graphic 1 Condition monitoring of pumps and systems

period. This systems offer protection from sudden breakdowns, increase the operational safety, make the saving of reserve pump possible, and contribute to saving costs (See figure 40 [24]).

Through electronic monitoring system it can be largely ensured that the pumps are operated in the scheduled operating state and disturbances are recognized and cleared in time. Indeed for more complex pump systems and special pumps, a monitoring system makes sense:

- Monitoring proof, why and when a pump breaks
- Measuring of electrical current and voltage, connections, temperature, pressure, flow rate, operation hours and gas content
- Measuring of the air quota (e.g. switching off at air aspiration of the pumps)
- Indicating of an error report, warning report at wear, flare or acoustics signal
- Examples of damage indication: damage of bearings, cavitation, leakage, delivery pressure too highly/too low, outlet end closed, inlet end closed
- Monitoring of several pumps (5, 10, 20 or the like) over PC perhaps as a service.

However, the application of the systems for the condition monitoring (CMS) contains more than only a monitoring system. This concept already starts at the damage avoidance. The next step, the identification of the fault is in the focus (see graphic 1). Concluding, by fault management the damage or the fault will be evaluated to avoid secondary damages. Also remote diagnostics by means of modem are possible.

3.4.1   Failure Prevention

A good planning of the pump system already helps preventively to avoid damages. This means that the layout of every component of the system like the pump itself must be done very exactly. If an exact complete analysis of the system was carried out, the probability of unexpected fault is very low.

Expected errors, such as repairs caused by wear out, are included in plans of this method of approach. By the installation of an overload protection device, the damage can be avoided immediately before occurrence.

3.4.2   Fault Detection

Also by good planning, disturbances cannot completely be excluded. To be able "to regulate" adequately, the system is monitored with the help of sensors. Changes of pressure, temperature, vibrations, flow rate and torque give information of measurement about an initiating harm.

3.4.3   Fault Management

The fault management serves to diagnose recurring disturbances immediately and for avoiding of secondary damages.

In addition, faults are evaluated after detection, and will be documented for being fixed in a matrix, according to a classification instruction.

The exact analysis of the causes helps to eliminate the error specifically and to avoid secondary damages by introduction of foresighted measures.

3.4.4   Pump Control 8

This system is a further development of the system "pump control", which is described under chapter 3.3.2. The electronics for the data acquisition of the operational parameters is situated in the clamp box (protection class IP 55). Through an RS 485 interface cable, the data are transmitted to a central interface unit. Per unit, 8 pumps are registered (see graphic 2 [24]). The system is expandable, for example to 20 pumps. Over a service adapter (RS 232), data are transmitted to the central control panel or service computer.

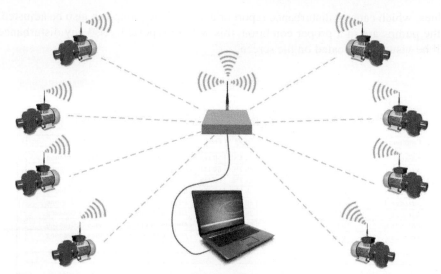

Graphic 2 Monitoring system pump control 8 monitors 8 pumps

The following operation parameters are to be registered (see figure 41 [24]):

- Operation hours
- Number of dry runs
- Number of motor overloads
- Number of power switching
- Pressure
- Flow rate
- Electric current/voltage
- Direction of rotation
- Phase failure

Figure 41 Monitoring system pump control is assembled on the pump instead of clamp box.

The monitoring parameters are electric current and temperature.

The set points and the permitted limiting values per pump to be monitored are menu-driven programmed. The current state data are indicated at the screen into tabular form, as well as the diagnosis and error indicator (see figure 42 [24]). Via the PC program, over the corresponding button, a respective pump can be turned on or off. Through the same interface connection (bi-directional) the set values and limiting values, which cause a disturbance report or a stop of the pump can also be adjusted. If the pumps are in a proper condition, this will be reported. Also any disturbance will be visually indicated on the screen.

Empfänger	Type	Zustand	Temp.	Freq.	Diagnose	Fehleranzeige
Adresse 0	001	Ein: 3.60 A	43.1°C	50 Hz	Alles in Ordnung	Alles in Ordnung
Adresse 1	002	Ein: 5.83 A	45.1°C	50 Hz	Alles in Ordnung	Alles in Ordnung
Adresse 2	003	Ein: 1.84 A	41.2°C	50 Hz	Alles in Ordnung	Alles in Ordnung
Adresse 3	004	Ein: 1.81 A	36.8°C	50 Hz	Alles in Ordnung	Alles in Ordnung
Adresse 4	005	Ein: 4.19 A	40.7°C	50 Hz	Alles in Ordnung	Alles in Ordnung
Adresse 5	006	Ein: 5.61 A	35.8°C	50 Hz	Alles in Ordnung	Alles in Ordnung
Adresse 6	007	Ein: 5.86 A	45.1°C	50 Hz	Alles in Ordnung	Alles in Ordnung
Adresse 7	008	Ein: 2.34 A	39.2°C	50 Hz	Alles in Ordnung	Alles in Ordnung

Figure 42 User interface with condition diagnosis

However, the monitoring can be carried out both, at the PC in the production hall both, but is also practicable as a teleservice (see graphic 3 [24]).

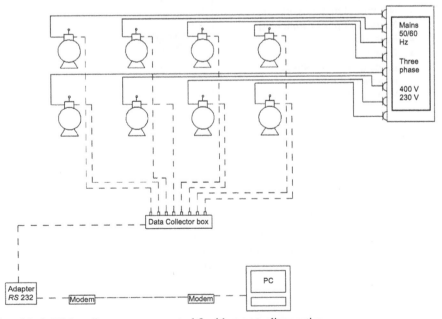

Graphic 3 Wiring diagram pump control 8 with remote diagnostics

3.4.5   Pump Monitoring

This diagnosis system makes it possible, to determinate a damage with the help of the vibration analysis.

Every type of pump has its own typical vibration characteristic. These vibrations change at different operating conditions (cavitation, dry run, closed slide, etc.) but also at defects at the pump, such as broken mechanical seal.

By the analysis of the rotary speed, a clear "vibration fingerprint" can be compiled for any pump. Through this, a reliable conclusion about the general condition of the pump is possible.

Here, the monitoring parameters are vibration and temperature. For the vibration measurement, a solid borne sound sensor, for the speed measurement, a speed sensor (laser sensor) is inserted. The measurand at the vibration measurement is the acceleration and is measured in g (9.81 m/s^2).

This system is a self-learning system. At the first take-off into operation, the actual values of the monitored pumps are identified and stored. For each pump, a characteristic is installed, disturbances are simulated or triggered. A set range of the parameters (green area) and a disturbance area (red area) are defined so that, later, the system immediately diagnoses the state, i.e. the type and dimension of the error.

A service employee can immediately diagnose which pump and in which current condition it is, by a signal of the system "normal condition" (green) or "error" (red).

The program consists of the following areas:

- Pump administration
- Set-up measuring channels
- Set-up graphic
- Measuring start
- Measuring storages
- Measuring loading

The user interface on the control panel computer with online diagrams forms itself as follows:

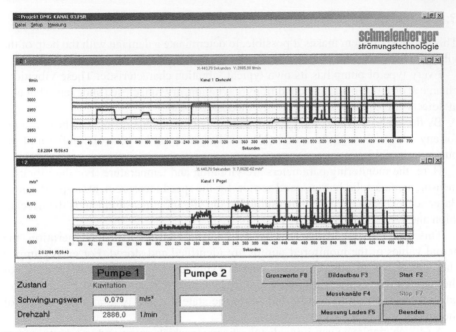

Figure 43 User interface with online diagram

At cavitation, speed and vibration are increased strongly.

Besides speed value and vibration value, "Cavitation" is shown as a disturbance type by means of a clear text message (see figure 43 [24]).

The different measurements as well as the evaluations are illustrated in the following diagrammatic form.

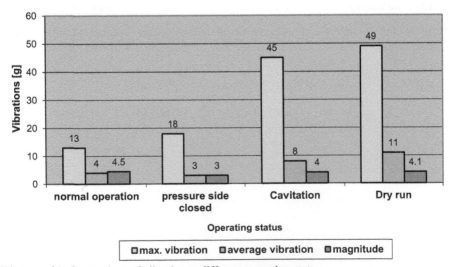

Diagram 21 Comparison of vibration at different operating states

In diagram 21 [24], the solid borne sound vibrations are shown as maximum value and as average value. The solid borne sound vibrations are measured in terms of g (acceleration), here, the highest value is 49 g in the operating state "dry run". The maximum values must not occur very frequent, it might be, that the "peak" arises only once during one measurement.

The Magnitude is the mean average value of the Fourier analysis, a mathematical method to analyze vibrations. With its help, the amplitude is shown to every frequency.

3.4.6   Contracting

To operate pump systems according to the contracting model includes, that the contracting party, the "contractor", runs the system or the pumps in authority and provide a trouble-free operation.

The model is based on the idea of the Scottish inventor James Watt (1736-1819) who offered at that time already the operation of a steam-driven engine as a service. This service was: installation of the machine and customer service for 5 years.

Contracting for the operation of pumps or pump systems means the provision of the pumps by the service company to periodic maintenance, repair and exchange of the pumps inclusively.

Therefore the pumps are not purchased but provided like the leasing proceedings against annual contracting charges. The operator of a plant saves costs for the investment, maintenance and repair as well as possible energy costs. The contracting charges, agreed on over several years by contract, cover these costs. The advantage for the operator of the pumps arises that he "buys trouble-free, optimized operation of the pumps" at long last and must not take care of anything. The contracting party who is just specialized in this trouble-free operation takes care about everything else.

The systems for the condition monitoring (CMS) find here their ideal application. However, the contractor also has great interest that the pumps run without interference on one hand, but also only few repairs occur, and the energy consumption is optimized. The operator or owner of a plant saves specialist staff who have to be provided by the contractor.

The opportunity of a teleservice system is obvious. The service employee of the contractor monitors externally, by use of modem from his office the systems of the customer. The repair can then be planned very exactly if required before the service employee has to drive to the customer. This can be regulated by a service contract.

Advantages for the operator of the system in detail:

- Everything from one hand
- Provision of operational pumps
- Reliable pump operation, rise of the operational safety
- Early diagnosis of critical operating states
- Wear out and life time forecast
- Repair or exchange of the pumps before damage

- By the less loss of production or unforeseeable costs
- Quick disturbance remedying
- Remote monitoring
- "Investment without money" (owner does not have to invest himself)

Through specialized mode of operation of the contractors offers the possibility for the pump operator to concentrate toward his core business like the operation or the production with the pump systems. The work on long failure analysis and repairs, analysis of harms and supervision of the maintenance schedules will be performed by the contractor. Loss of production then should become a rareness.

3.5 Flow Optimization

According to law of energy conservation of the Swiss Scientist Daniel Bernoulli, no energy get's lost in the flow either. The sum of all energies is constant in every point in the system. Friction energy, in the piping system and in the pump, however, can reach a remarkable quota, determined by the system. This can lead to considerable flow losses at long last. The influence parameters, which affect directly, the value of the head loss:

- Cross-section of the pipe
- Length of the pipe
- Inner roughness of the pipe
- Viscosity and density of the delivery fluid
- Type of flow (laminar and turbulent)

The laminar flow also known as a layer flow causes little decrease in pressure comparatively and appears especially also in straight pipes. For very rough inside walls of the pipes and increased flow velocity, the laminar flow turns to the more lossy turbulent flow. It appears especially into fittings like gate valves, valves, tube narrowings or tube expansions. If possible, areas in the system, which generate a turbulent flow, should therefore be avoided.

The losses in the pump can be summarized and appear in the following places:

- Suction pipe
- Inlet edge of the impeller
- Backs of impeller vane
- Backs of impeller
- Pump discharge nozzle

The impeller should be optimized accordingly. The suction pipe should always be larger than the pump discharge nozzle.

3.5.1 Losses of Pump Components

On surfaces of impellers and spiral casing roughness should be avoided. According to the theory of boundary layer, it produces turbulences and therefore causes losses by a turbulent component of the flow. More opportunities for optimization are:

• Use of ball bearings with little friction
• Reduction of the gap losses between impeller and spiral casing (volute)
• Additional smoothing of the surface of impeller and spiral casing

3.5.2   Losses in Pipes, Components and Fittings

If there is a laminar flow in the tube, the flow loss in the straight tubes remains relative in limits.

However, to minimize the flow losses in the pipeline system, a sufficiently wide diameter of the tube should be chosen (see table 11 [24]). But the dependency of the flow losses of the type and composition of pipes, elements and fittings must be highly taken into account (see figure 44 [24]). The following items are having particularly a great influence on the flow losses:

• Elbow joint
• 90° bow
• Stepwise narrowing
• edged inlet
• Slide
• Throttle-/ gate valve

Some parameters are listed exemplarily in the following:

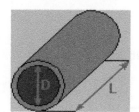

Boundary conditions: turbulent flow of water, flow: 48 m³/h, roughness value: 0,1 mm

Figure 44 PVC-pipe

Table 11 Head Loss According to Flow Velocity and Pipe Diameter

Tube, Length: 30 m, PVC Hard, DIN 19532 (Nominal Diameter)	Head Loss	Flow Velocity
DN 80	**0.263 bar**	2.562 m/s
DN 100	0.095 bar	1.718 m/s
DN 150	**0.014 bar**	0.812 m/s

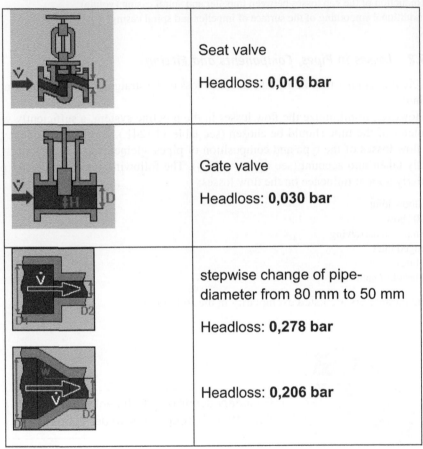

Figure 45 Head loss at nominal diameter DN 80 and a flow velocity of 2.562 m/s

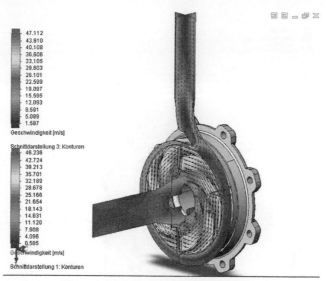

Figure 46 Flow simulation of the system [24]

It can be summarized, that any change of the pipe width or change of fluid-flow-direction causes losses (see figure 45 [24]).

A specific choice of the components helps to reduce losses and therefore to save energy and costs.

3.5.3   Optimization by Flow Simulation

Both, the flow of the delivery fluid and the damages which result from abrasive media inside the pump can be predetermined by flow simulation.

To be able, to analyze the flow of delivered medium into pump and system - more accurately, the flow course can be calculated by appropriate CFD software (CFD = Computational Fluid Dynamics). Therefore, the exact geometric data for pump geometry and to the system as well as the flow data like flow rate, pressure, etc. must be available. For the computation, the geometry data of the pump system are processed with the software. A networking (grid) is then carried out as a preparation for the calculation of the flow [28].

Each element, fitting or section of the pump which should be calculated, has to be netted with a mesh.

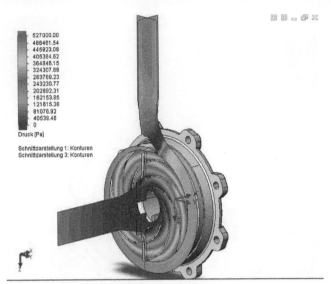

Figure 47 Head loss in the system [24]

The CAD data are used directly to record material and flow cavities. With the help of a mathematical and physical model for laminar, turbulent and transition flows, the relevant modelings are established. From this modelings, different operating states then can be simulated.

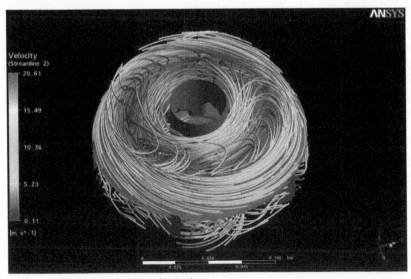

Figure 48 Liquid filament, course of velocity

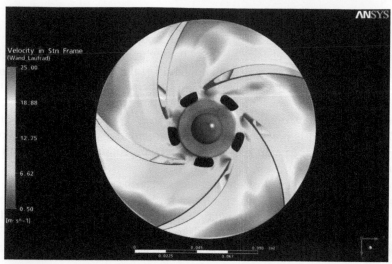

Figure 49 Absolute velocity on the impeller surface

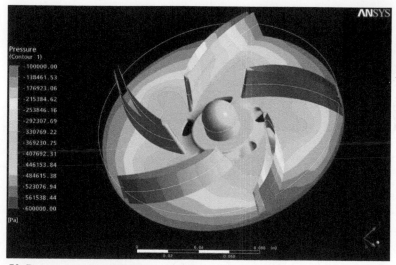

Figure 50 Pressure course within the impeller

A flow simulation is advisable to estimate the flow conditions into pump and system. By the simulation of different operation states, the most efficient mode of operation can be investigated (see figure 48,49,50 [24]). The expected damages also can be shown very well by flow simulation.

3.5.4   Flow and Wear out Simulation

For the recording of the erosive strain of the component surfaces, numeric flow simulations (CFD) can be elaborated, too. Thereby the transportation of particles in the fluid can be modeled. By transient simulation of the flow conditions, the places where material erosion at the impeller occurs, are being evaluated. Through integration of the particle-wall-interaction during the running time of the CFD calculation, erosion rates can be determined qualitatively at every point on the component surface. Through this, it is possible to diagnose particularly wear out charged places during the lay-out of the pump, and take into account both material technically and constructively.

To examine the ductile deformation processes for the wear out simulation, finite element nets are established. This enables to investigate the strain, which predominates in the material on micro structure level. These nets are then provided with realistic boundary conditions, which permit to calculate the local deformation when outer loads impact. Conclusions on the potentials for the material optimization can be derived out.

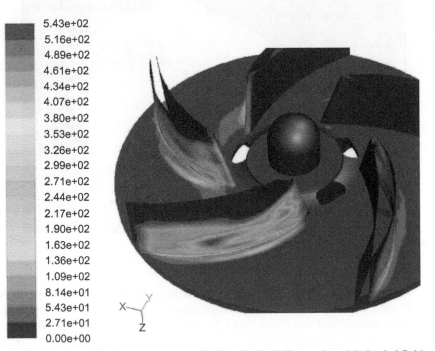

Figure 51 Number of particle elements on the impeller by delivery of particle loaded fluids.

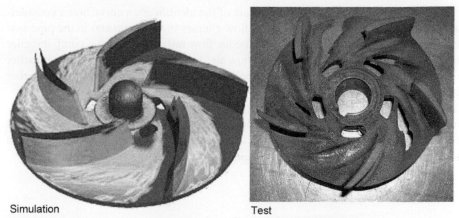

Simulation   Test

Figure 52 and 53 The wear out simulation - spread of the impact velocity of the Particle

In the above shown pictures (figure 51,52,53 [24]), it can be seen clearly, that the particles are captured by the vanes of the impeller and then glide along the top side of the vane. Through evaluation of the impact angles, the impact velocity and the prevailing flow conditions it is possible to identify the damage images, which corresponds to damage mechanisms arising in the test facility.

At the flow and wear out simulation the following findings arises:

* The speed vectors show the highest speed balance between impeller and medium at the inlet edges of the impeller.
* Turbulences, which affect a relatively big area of the vane channel form at the impeller vane back.
* Very strong signs of wear out occur in the area of these turbulences.

The examinations showed, that wear out is mainly dependent on the kind of erosion of the influence parameters particle speed, the impact angle and the ductility of the material.

In the case of a flat impact angle of the particles, it comes to a ductile deformation. When the particles hit vertically on the surface, it comes to "crater formation" in the meeting point.

With approximation formulas for the angular dependence of the material erosion, for the uncoated material a maximum of the erosion was calculated at 12 - 13° impact angle. On hard surfaces (coated), the maximum erosion appears at a steep impact angle of 80-90°. The images of damages on individual components are almost congruent with the results of the flow simulation. [17]

Summary of wear out simulation

Through the wear out tests and examinations it could be investigated, that the surface hardness alone is not sufficient as factor for wear out avoidance. The numeric examinations show potentials for the improvement on the characteristics of the surfaces of the pump components.

The simulations show very well the data of the identification curve under consideration of model-like simplifications. The flow courses in the pump and in the pipe system can be well described. By abrasive media, flow turbulences on the pump impeller produce wear out. The speed vectors show the highest difference between impeller and medium at the inlet edges of the impeller. The higher the flow velocity, the higher the wear or the material erosion.

The tests show clearly that the images of damage of the individual components agree very well with the results of the flow simulation. The places with an increased flow velocity show the strongest signs of wear. The wear simulation is therefore a good opportunity to predetermine damages being expected, by an abrasive wear.

4 Measures at Components for Reduction of Wear Out

Without consideration, which pump components then are damaged at long last, has the kind of operation a great influence on the condition of the system, if it is free of damage and long lasting. Since there is a direct relationship between speed, wear and noise level, it is necessary to reduce the flow velocity of the fluid in the pump system for the optimization of the process. But this must be done use-related to the application. The lower the flow velocity, the lower the wear and the noise level. Therefore, if possible pumps with a lower speed, e.g. 4-pole-motors with speeds of 1.450 rpm (rounds per minute) or frequency converters should be used for variable speed drive (VSP).

Wear reducing protective measures and surface conditioning can be carried through at the following components:

- Impeller
- Spiral casing
- Pressure lid
- Bearing
- Pipelines
- Mechanical seals (jamming chamber etc.)
- Further components

If necessary, also constructive measures, such as wear decreasing inserts make sense.

4.1 Anti-Corrosion Protection

Different coating systems as corrosion protective layers are suitable. Organic, inorganic or non-metallic diffusional coatings. Before the respective layer however, can be applicated, it requires a pretreatment. The frequency range suffices from cleaning-dealt by a mechanical cleaning up to thermal processes.

4.1.1 Procedure of the Surface Preparation

Brushing

The simplest opportunity of the mechanical surface preparation is brushing. Particles which release easily and on the surface are removed. Either nature bristles or bristles come either from synthetic material or from metal into application.

Damages on Pumps and Systems. http://dx.doi.org/10.1016/B978-0-444-63366-8.00004-8

Blasting

An abrasive (sand or glass beads) is pneumatically, hydraulically or mechanically (spinner) taken to the impact on the metal surface with a high pressure. Through this, also the particles which cannot be removed by brushing, will loosen itself.

Grinding

With the help of granular abrasive compounds or by steel wool, the metal surface will be implemented.

Shaving

The surface is prepared by shaving with a hardened steel blade.

Cleaning with Wire Needle

With the help of a wire needle, air-atomizing spray gun pollutions can particularly be removed from edges and corners.

Flame Cleaning

With a concentric jet burner, sedentary substances like rust or little more tightly tinder can be removed by a short-time blasting.

Bright Annealing

Thin oxide films are removed at high temperatures by deoxidizing gasses.

4.1.2   Classification of the Corrosion Protecting Layers

Organic Layers

Coating substances made of pigments, synthetic materials, or bituminous substances are applied on the metal surface.

Inorganic Layers

Paint-like coating substances (e.g. zinc ethyl silicate), filler material and binder based on ethyl- or alkali silicate are applied on the metal surface.

Ceramic Coatings

By thermal spraying, the ceramic layer, consisting of inorganic crystals is applied on the metal surface.

Condensation Technique

One distinguishes between physical and chemical procedures. By physical PVD process (physical vapour deposition) the layer is established on the metal surface to be protected by means of a non metallic coating substance which vaporizes due to condensation in the vacuum.

By chemical CVD process (chemical vapour deposition) the layer is produced by a surface catalytic reaction with gaseous compound of the coating substance on the metal surface.

Oxide Coatings

Thermally produced, by oxidation of steel in hot air or in salt-melt, thin, blue oxide coatings arises. In alkaline saline solutions, the layers are deep brown to black.

Non Metallic Diffusional Coatings

Nitriding protective layers are produced by the metal being glowed in nitrogen giving chemicals. Boron layers are produced with substances giving boron (powder-, granular- or paste-shaped).

(Source: DIN 50902:1994-07)

4.1.3 Material Selection

A damage can already be minimized or excluded preventively, depending on application case and the wear to be expected. If corrosion is expected, gray cast iron should be excluded as material. In the context of given boundaries, high-grade steel (1.4571, 1.4539) can be used. At a value of 4 500 mg/l CL-ions , high-grade steel 1.4571 should be used, at a value of 18 000 mg/l CL-ions , high-grade steel 1.4539/1.4462 (cast) or bronze (2.1050) should be used. At a value of 30 000 mg/l GL-ions, AL bronze (2.0980/2.0966) should used.

Provided that the strength requirements permit it, the safest version to prevent corrosion, is by using pumps or pump parts made of synthetic/plastic materials.

4.1.4 Plastic Coatings

Plastic coatings can be conditionally used against wear out. As anti-corrosion protection or also among little abrasive delivery fluids, plastic coatings are suitable.

The choice of a coating system means not only the choice of the layer itself, but the choice of a suitable combination from base material to coating, layer thicknesses and material transitions.

By coatings two groups of layer systems are distinguished in principle:

- Reaction layers – arise by influencing for the surface layer – e.g. by heat, diffusion or implantation of alloying elements
- Plating layers - arise by application or sedimentation of elements

Plastic coatings belong to the plating layers. By little abrasive media, plastic coatings are very well suitable for the smoothing of the surface and therefore for the reduction of flow losses (see figure 54 and 55 [24]). Tests have shown that the efficiency factor of the pump could be increased by up to 9 %.

Figure 54 and 55 Impeller and spiral casing with plastic coating [24]

In the table 12 [24], the characteristics of 2 materials are exemplarily listed, which are suitable as plastic coating.

Table 12 Material Parameters of Selected Synthetic Materials (Plastics)

Material	PFA (Perfluoralkoxylalkan)	ETFE (Ethylene-Tetrafluorethylen)
Water absorption (up to the saturation)	low: < 0,03%	low: 0,03%
Use temperature	– 200 to + 260 °C	– 100 to + 150 °C
Other characteristics	anti-adhesive behavior	anti-adhesive behavior
Layer thickness	0,4 – 0,6 mm	0,4 – 0,6 mm

Worn out pump spare-parts, which have been seriously damaged by corrosion, erosion or cavitation can be repaired again by plastic coatings. The damaged parts are subjected to a thorough pretreatment before the coating process is conducted.

Depending on the degree of damage, by overlaying welding, the necessary strength is given back to the parts again. The surface is afterwards sand-blasted (see figure 56 [U & Z]). After an undercoating layer on a metallic basis (see figure 57 [U & Z]), the final coating is carried out with a ceramic-strengthened epoxy layer (see figure 58 [U & Z]).

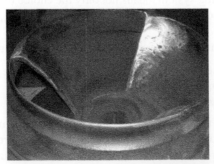

Figure 56 Impeller after overlaying welding and sand-blasting [U & Z]

Figure 57 Impeller after primer repairs [U & Z]

Figure 58 Impeller with 2 C (2 components) ceramic epoxy coating [U & Z]

The pump spare-parts are almost as good as new after the repair and are fit for use for another operating cycle.

Plastic coatings are not suitable by strongly abrasive medium, therefore, so-called "hard layers" can be applied here.

4.1.5 Anti-Corrosion Protection of Stainless Steel

The corrosion resistance of stainless steel is determined by the type of steel and the medium which comes into contact with the surface. A corrosive atmosphere, such as saltwater or chemicals, is able to destroy the stainless steel. The condition of the surface plays an important role. Untreated, sharpened or polished, the corrosion resistance can be improved.

In principle, the passive layer forming on a natural way protects the steel from corrosion. The chromium in the steel reacts with the oxygen of the ambience and forms chromium oxide as a passive layer. Besides chromium oxides, passive layers also contain iron oxides. The higher the chromium content, the higher is the corrosion resistance.

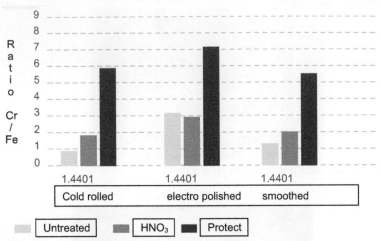

Diagram 22 Chrome-/ iron ratio of different surfaces and types of treatment of the stainless steel 1.4401 [20]

By using a chemical process, the passive layer can be strengthened against corrosion. Thereby, the surface is enriched with chromium and follow-up treated under heat supply (140° C to 220° C temperature) aftertreated.

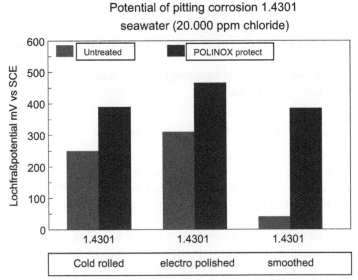

Diagram 23 Pitting corrosion resistance in dependence of the surface composition [20]

The more sophisticated is the surface, the more resistant is the passive layers against destruction (see diagram 22 [20]). In the diagram 23 [20], the pitting resistance of the stainless steel 1.4301 is drawn at different surfaces. The electro-polished surface with a follow-up treatment shows the best resistance. Therefore, the passive layer is decisive for the corrosion resistance on the stainless steel surface.

4.1.6   Other Options

As an additional option to hold on against corrosion, is the variety to manufacture the pump spare-parts made of full plastic. Or if it is necessarily, a stainless steel frame work may be put around the pump case in addition, as shown in figure 59 and figure 60 [24].

Figure 59 Plastic pump with stainless steel shell

Figure 60 Plastic pump with metal frame

4.1.7 Anti-Corrosion Protection of Mechanical Seals

Mechanical seals are very differently suitable for corrosive media.

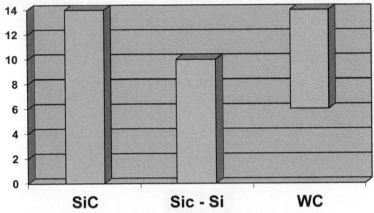

Diagram 24 Comparison of the corrosion resistance SiC, WC, Sic-Si

From this comparison of the material characteristics it is obvious, that silicon carbide (SIC) is suited best at applications with corrosion risk. Better than wolfram carbide (WC) (see diagram 24 [24]).

4.2 Abrasion Protection

To be able to decide on the right protective measures, sufficient information must be available about the mode and the dimension of the signs of wear out. Visual assessment, i.e. circumstances permitting the implementation of a test should be undertaken. Sufficient information also should be available about the kind and quota of the solid matters being in delivery medium. A flow analysis should be

Figure 61 Damages by corrosion and abrasion to the spiral casing

carried out especially by more complicated damages. Only after that, it should be decided, whether constructive measures are necessary or a surface protective measure makes sense.

By very extreme, abrasive demands (see figure 61 [24]), it is recommended to think about the implementation of a wear out monitoring concept. Which monitoring system is then used or suitable - stationarily permanently, mobile or with a cyclical measuring - such decision can be done only after a thorough cost-benefit analysis.

4.2.1   Wear Out Analysis

In every case, the analysis should start with the check of the pump data, using the completion certificate. Many damages are already identified in the system scheme, so that this scheme should be taken into consideration. If no mistakes are recognizable here, the next step is the damage investigation with the interpretation of the damage image.

The examination of the damage image, done by a documentation with photos/pictures and sketches, makes the work very much easier. It is advisable to carry through, a damage examination in accordance with the guideline VDI 3822. Main emphases are the three steps of the damage analysis: damage description, baseline study and damage hypothesis. Individual examinations, to be carried out for the evidence of the damage hypothesis, can be done in addition if necessary, too (see figure 62 [24]).

The Most Important Terms of the Damage Analysis to VDI 3822 are Compiled Together as Follows:

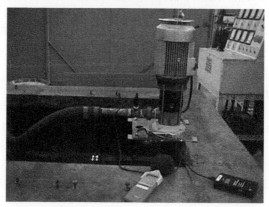

Figure 62 Failure analysis at the testing basin

* Damage - changes to single components by which the scheduled function is affected or made inevitable.
* Damage type - name of the damage (e.g. breakage, abrasion, corrosion)
* Component – affected component or fragment of a component, concerned by the damage
* Damage image – outsides condition of the damaged component
* The damage place - place of harm on damaged component

- Damage appearance - distinctive features (characteristics) of the damage type
- Damage sequence of operations - time development of damage (until the breakage)

The description of the damage image should indeed be performed by a documentation of damages with the help of photograph, outlines and number data and possible measurements at the best one. All conspicuous features should be put on record. Damaged components or the pump areas concerned must be inspected.

Dimensions, constructive and production technical details, as well as the installation data and the specific details of the strain should be recorded. Appearances, general situation and source of deformations, rips, cracks, breaks, and signs of corrosion and wear out should be well determined and documented exactly. Material characteristics of surface give additional helpful information about damages.

4.2.2   Flow Analysis

If the damage by the abrasion is considerable and the reasons are not very plausibly to comprehend, it is recommend to carry out a flow analysis, by numeric flow simulations, using CFD (Computational Fluid Dynamics). Many parameters as possible and information about the boundary conditions should be available.

Especially about the delivery medium (fluid) should the following parameters be known: viscosity, quota of solids, temperature, flow rate, delivery pressure, flow velocity and possible the pH-factor.

Through the flow simulation, the data of the characteristic curves are indicated, recognizing model-like simplifications. The flow courses, flow lines and formation of turbulences at impeller vanes and spiral casings can be identified. On those areas, where huge wear out occurs and material erosion takes place can be clearly observed.

After comparison of the results of simulation with the actual damages, then it can be decided, where and at which components measures make sense.

4.3   Solutions by Design

Preventively, wear out can through respective measures be compensated and even avoided.

Shaft Protection Sleeve

The shaft is provided with a protective sleeve in the abrasion-risky places.

Relief Wells

Relief wells at the impeller serve the reduction of the axial load and the protection of the bearings.

Wear out Plate

Wear out plates are exchangeably and assembled behind the impeller, for the protection of the spiral casing.

Ceramic Inserts

By very hard solid matters, remedial action can be taken by installing ceramic inserts on the very stressed places.

Monitoring by Means of Sensor

Monitoring by means of an abrasion sensor offers the opportunity of the wear out to be indicated by a signal.

4.3.1 Wear Out Reducing Inserts

To extend the life time of the pump cases, a quite simple and economical measure offers its services with a wear plate in the pump case. Since the pump case does not have to be taken off with the piping in the claim, for the exchange of the wear out plate, the effort for replacement is relative comprehensible. Price comparison with a new pump case is not necessary.

Figure 63 Wear plate in the pump case (before damage)

Figure 64 Abrasion at the screw bores (after damage)

By comparison of the two photos (see figure 63 and figure 64 [24]), it is recognizable that the abrasive material removal starts at the edges of the screw bores. Before the complete, quite thick wear-out plate is worn out, the screw heads fall down most likely. If the delivery fluid contains very hard wear particles, wear protection - made of ceramics - for system components and pipes is recommended (see figure 65 [11]).

Figure 65 Different ceramic Inlets [11]

Such inlets are available with very different wall thicknesses, sizes and geometries. For large pipes they can be fitted in very easily. By smaller pipe diameters, limits are set out on wall thicknesses. However, special designs are usually practicable.

4.3.2   Monitoring by Means of Sensors

The material abrasion on the impeller can be detected by a decreasing power output of the pump, while the abrasion on the pump case is noticed only at the leakage of the delivery fluid. This is a critical factor on pumps for dry installation, since the leaking fluid can then run out to the surrounding area. To recognize this wear-out earlier and be able to start appropriate maintenance measures, it is very suitable to use a system with wear-out sensor (see figure 66, 67 and 68 [24]).

Figure 66-68 Installation situation of the sensor

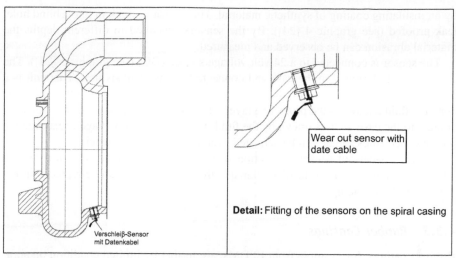

Wear out sensor with date cable

Detail: Fitting of the sensors on the spiral casing

Verschleiß-Sensor
mit Datenkabel

Graphic 4 Abrasion sensor in the pump spiral casing (volute)

The wall thickness of pump case components should be monitored with the help of sensors. The wall thickness should be monitored on such spots where material removal by the abrasive fluid is very huge. Due to material removal, arises the risk of leakage through the leak on spiral casing on the surrounding. Here a system with a wear sensor is useable.

4.3.2.1 Functional Description

The wear sensor is installed at case areas which are affected particularly by the abrasion. At this sensor principle, the conductivity of the fluid is utilized. The sensor, whose core consists of conductive material has a cylindrical form which is surrounded

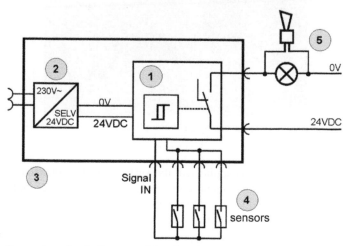

Graphic 5 Connection circuit diagram of the abrasion sensor

by an insulating coating of synthetic material. The sensor is stuck in into a blind hole, leak-proofed (see graphic 4 [24]). By the sensors, installed in different depths the material abrasion can be observed and measured.

The sensor is connected to a 24 volt voltage source (DC) (see graphic 5 [24]). The ground line is fastened to the case and connected to the analysis unit. This unit is a threshold switch, whose shift point is adjusted to a maximum sensitivity of 1 V. The abrasive fluid excavates the protective layer on contact with the sensor. The electrical contact to the case, then is drawn by the fluid. If the switching voltage is exceeded, a flare and acoustics signal will be activated over a potential-free switch.

If such a wear-out sensor is positioned very well, foresighted measures can be taken. Before the wear-out leads to damage, the spiral casing can be replaced in the context of the servicing.

4.3.3 Rubber Coatings

Rubber coatings, for example from EPDMs, are usable by limited condition on abrasive delivery fluids (see figure 69 [24]).

EPDM (ethylene propylene diene monomer) as a type of synthetic rubber, is an elastomer with a wide range of applications.

Since the rubber coating is softer than the metal particle, the rubber, as an elastic material is resilient, when the particles are colliding. The particles bounce off and are further transported with the delivery medium.

Figure 69 Rubber coating of spiral casing and back plate

Before the rubber coating is applicated, the surface must be cleaned very well, so that the adhesive strength on the metal gets firmly enough. A primer (adhesive layer) must possibly be used. The thickness of the rubber coating should be max. 4 mm. The solid matter quota in the fluid should not exceed 15%.

It is a disadvantage, that the pump components must be strongly surface finished, before the rubber coating, so that the layer thickness of the rubber coating can be compensated.

Depending on strain, however, the wear-out will also not fail to appear sometimes. After certain time, nevertheless small rubber particles are excavated and small cracks also might arise. The service life of a pump can absolutely be doubled, depending on stress also more. After first damages, however, the rubber coating might be worn out and excavated partly.

4.3.4 Wear Resisting Materials of Cast Iron

By abrasive, corrosive strain, white cast iron is very often used. This material also indicated as chilled casting or hard cast is very much wear-out resistant. The exceptional feature of this material includes that the carbon content is chemically bonded as carbide. The fracture surface is white or silverly, unlike the gray of the cast iron. There are different kinds of hard cast, depending on alloying quotas and alloying degree. Depending on the structure by unalloyed or low alloyed sorts, chromium carbide, molybdenum carbide, niobium carbide or vanadium carbide arise. The hardness of the carbides reaches from 800 Vickers (HV) by cementite, 1 600 HV by chrome carbide up to 2 800 HV by vanadium carbide (see table 13).

Table 13 Hardness Values of the Cast Iron Materials in Vickers (HV)

Material	Hardness in Vickers [HV]
Cementite	800 HV
Chrome carbide	1 600 HV
Vanadium carbide	2 800 HV
Cast iron	210 HV

A hard cast, where nickel and chrome as alloying components are contained proportionally 2:1, is known under the trade name "Ni hard" (e.g. Ni-hard 1 or Ni-hard 4). The qualities can be varied by the change of the carbon content [38]. Although, the chrome {chromium} content is partly very high (> 20%), the hard cast it not very corrosion-resistant The reason is, that the main quota of the chrome is bound in the carbides and the content of other alloying elements, such as nickel or molybdenum, is not sufficiently high. Nonetheless the corrosion resistance is strongly dependent on the chemical composition of the delivery fluid.

Such materials can be applied well by strongly abrasive media, such as sand-/ water mixtures. However, since the alloying costs are relatively high, this application will be suitable mainly for very special solutions. In addition, the components made from hard cast require special cast tools.

A post-machining (trimming, threading) of the castings is only possible with special tools.

For example, a power matching of the pump by changing the impeller diameter by trimming is complicated.

4.4   Surface Reimbursement

This is not an upgrading by a galvanic procedure, but a specific "hardness increase" of the surface layer which causes an increase of the strength against abrasion. As already described in chapter 4.2.1, the selection of the wear protective layer is based on three steps of the systematic failure analysis: Damage description, fact-finding and damage hypothesis. After an analysis and exact knowledge of the damage image or the wear-out appearance it makes sense to fix the requirements on the layer sub-stratum system. On one hand the tribological requirements as well as constructive and manufacturing-related requirements have to be fixed to conclude finally the material requirements. A selection of suitable wear-out protective layers can then be carried out.

4.4.1   Hard Layers

As already described in chapter 4.2.1, the choice of a coating system is not only the choice of the layer itself, but it means the choice of a suitable combination of base material, coating, layer thicknesses and material transitions (see graphic 6 [31]). The hard layers belong to the reaction layers.

The reaction layers arise from influence to the surface layer e.g. through heat, diffusion or implantation of alloying elements.

"Soft" materials can be hardened by a special surface termination in the surface layer. Consequently through their greater resistance to abrasive media, the service life of the pump can be increased considerably.

The adhesion of the layer on the cast iron surface is a problem by different sorts of coating processes. By an amplified sandblasting or glass bead blasting, the layer must be prepared for the coating process. By thin layers an additional smoothing still must be carried out to be able to obtain a satisfactory result.

The laser surface cladding is suited very well under the "high-build" ones, since the thickness of the layer definitely can be > 1 mm.

By the nitriding and nitrocarburisation the friction coefficient and the adhesion inclination is reduced, on the other hand increases the abrasion resistance and the strength against fatigue by change deformations [36].

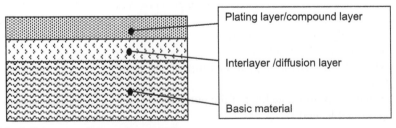

Graphic 6 Substratum layer structure [31]

4.4.1.1 Nitro-Carburisation

The nitro-carburisation implies a process of nitriding (gas or saline bath) which means that from the nitriding material, both nitrogen and carbon are diffusing into the surface of the component.

The item is exposed to a carbonous and nitrogenous surroundings atmosphere by a temperature of 480 to 590° C. By the nitrogen and the carbon, a change of the chemical composition and the structure takes place in the surface layer.

4.4.1.2 Plasma-Nitriding

The process of nitriding takes place in the plasma (electrically, conductive gas) in a vacuum stove at 350 to 600° C. The nitrogen from the nitriding material diffuses into the surface and causes a change of the chemical composition and the structure in the surface layer (see figure 70 [24]). In this context, a very low deformation and an only insignificant size alteration arises through this process (see figure 71 [24]).

Therefore, this process is for thin-wall components more suitable than the nitro-carburisation.

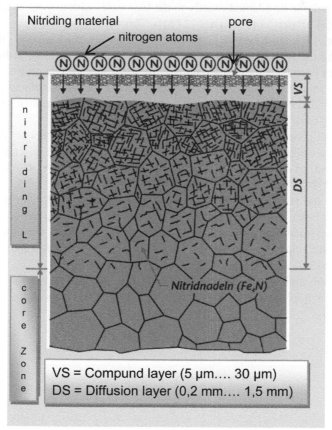

Figure 70 Nitriding layer and core zone after the plasma-nitriding process [24]

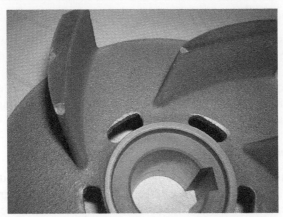

Figure 71 Impeller with nitriding layer after the plasma-nitriding process

4.4.1.3   Laser Surface Cladding

During the laser surface cladding, material is locally and precisely applied on to the surface.

The material can be adapted exactly to the load case, with respect to hardness and the mechanical properties (see figure 72 [24]). This layer has a hardness of 900 Vickers (HV), thereby it has a very high stability and causes by the high layer thickness of 1 mm a rise of the service life. An adjustment of geometry is necessary, though, since the hydraulics is changed (size alteration) by the laser surface cladding.

Due to the relatively thick layer and the special mode of the manufacturing method, wear-out plate and impeller can be coated - according to the requirements of the applications - up to max. 1000 µm (1 mm) layer thickness.

The low, but concentrated heat input guarantees very low deformations. The principle of laser surface cladding or plating consists in building up powdered material by welding on a component, so that tightly adhesive and increased marks arise (raised welding bead) (see figure 73 [24]).

Therefore, the material of the workpiece will be partly fused, with the help of a focussed laser beam in the focal spot. At the same time, additional material is brought into the molten puddle and also melted on. After the solidification, the additional material is tightly sticking on the workpiece. This process is happening fast and continuously (see figure 74 [24]).

By the combination of individual coating tracks, flat and multilayer coatings can be realized, such as they are needed for the repairs and the wear-out protection (see figure 75 [24]).

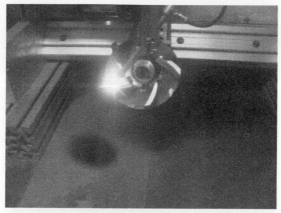

Figure 72 Laser surface cladding on the impeller

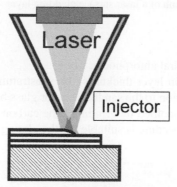

Figure 73 Principle of process

Figure 74 Impeller with laser surface cladded vanes

Layer thickness: 1000 μm
Hardness: 900 HV

Figure 75 Structure micrograph of a laser surface cladding layer

4.4.1.4 Carbon Layers

Carbon layer ta-C (tetrahedral amorphous carbon):
Due to the relatively thin layer thickness, the substratum must be preconditioned
before the coating. In addition to the sandblasting or glass bead radiation, the component surfaces should be smoothed by grinding. The carbon layer has a hardness of 4
500 Vickers (HV). This procedure is suitable for the plating of the impeller.

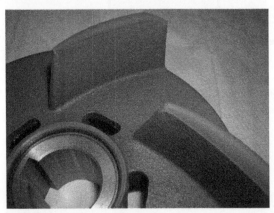

Figure 76 Impeller with amorphous carbon layer

The process: Laser controlled vacuum arc plating (laser arco)
The laser Arc plating principle combines the advantages of the processes VAD (Vacuum Arc Deposition) and PLD (Pulsed Laser Deposition). By the use of an economical

pulse laser, a temporal and local control of the arc discharge gets possible (see figure 76 [24]). With the help of a laser pulse, an arc with a very strong electrical current, with a peak current of about 1.8 kA (kilo Ampere) is sparked, whereby the duration is limited to 130 micro-seconds (µs).

The amorphous carbon layer is vapoured in the PVD process by graphite vaporizing.

The laser Arc module can be docked at every customary plating system (see structure micrograph figure 77 [24]).

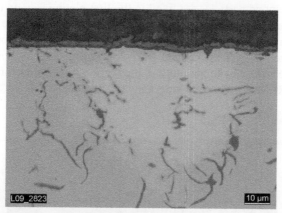

Figure 77 Structure micrograph of the ta-C-layer

4.4.1.5 Diamond Similar Layers

Duplex Layer

The very hard silicon carbide (SiC) is contained in this layer, whereby a layer hardness of 2 600 HV (Vickers) is achieved.

The layer thickness can be applied up to 100 µm. This layer is suitable for spiral casings and impeller (see figure 78 [24]).

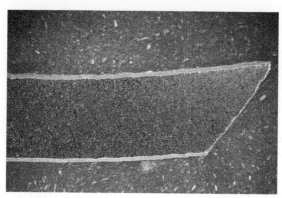

Figure 78 Micrograph of the duplex layer

Characteristics of the layer:

* Diamond particle sizes 1 - 45 μm
* Typical particle sizes: 1-3 / 8-12 / 20-30 μm
* Particle size: 1-3 μm
* Enclave volume: 25 - 40 %
* Layer thickness: up to 100 μm

Plating process:

The rough diamond particles are fixed in the first plating step on the surface. The free space between the rough diamond particles is filled out in the second plating step with another dispersion plating up to the desired embedding quota (see figure 79 "surface structure" [24]). This second dispersion plating contains considerably finer diamond particles. By the duplex plating, the wear-out resistance of the chemically nickel - matrix REM increases considerably.

Duplex layer:

Figure 79 Surface structure chemical-nickel-diamond (particle size 40 μm)

4.4.1.6   Summarizing and Evaluation of the Plating Processes

Different plating processes are suitable for the protection against abrasion (see table 14). Depending on application and solid matter (chips, grinding dust, sand, very hard solid matters), it must be decided about the process to be used and the layer thickness. A selective plating, especially by laser surface cladding, is very well possible, depending on application.

Table 14 Hardness and E-modulus (Elastic Modulus) after the Plating

Material	E-Modulus [N/mm^2]	Hardness [HV]
Cast iron, without layer	100 000	210
Cast iron + nitrocarburisation	110 000	600 – 660
Cast iron + plasma-nitriding	110 000	1160 – 1200
Cast iron + laser surface cladding	210 000	900
Cast iron + ta-C (carbon layer), (Diamor)	300 000	4 500
Cast iron + chemical-nickel-dispersion plating with diamond particles (duplex)	145 000	2 600

The production procedures are compared in the following evaluation matrix to give some basic information for the decision on the layer selections (see table 15).

Table 15 Production Processes

Layer:	ta-C	Duplex	Laser Surface Cladding
Layer thickness	10 μm	50 μm	1000 μm
Production processes	Vacuum arc plating (laser-arco) with amorphous carbon Electrical power input: > 20 KW	Fixing of the diamond particles on the surface and replenishment of the free space with dispersion plating Electrical power input: > 30 KW	Building up with weld of powder-shaped material. Electrical power input: < 10 KW
Process time for one impeller	4-5 h	2 h	5 minutes
Process-temperature	< 180° C	90° C	1 300° C
Tempering follows		350° C	

In the context of the predictive maintenance, a load profile can be created for pumps, using the wear-out data and the details on flow speed and operation hours. A flow simulation gives information about the strongly stressed sectors inside the pump.

Especially for special applications, an appraisal can then be made, about the service life of the pump to be expected.

An increase of the service life justifies therefore this additional costs arising for the plating, because through this the costs for the loss of production can be reduced enormously.

4.5   Special Designs

To minimize the damages at particularly wear-out afflicted processes and to increase the service life of the pumps to an acceptable dimension, different special designs are practicable in addition to the platings.

4.5.1   Pump with Cutting Unit

Among other things, pumps are also used in the field of metal cutting machine tools for delivering of refrigerant fluids, polluted with metal chips. The metal chips very often forms balls which hinder the delivery of the refrigerant fluid reliably. For saving expenses, steel wool shredders (steel wool breakers) or lifting systems, which are expensive and effortful are frequently renounced. For this reason it makes sense to combine pumps with cutting tools. Therefore, a pre-shredder gets preconnected on the pump shaft, mounted in front of the real impeller.

By different pumps on the market, pumps with cutting edge and pre-shredders are arranged inside the pump case. The pre-shredder is surrounded by the suction pipe of the pump so that the crushing of the steel wool balls to small pieces is happening in the pump case. Nonetheless there is a risk of the blockage, if the scouring with steel wool balls are too big or the chips are not cut small enough.

Other systems consist of a pump and a separate steel wool breaker (crusher). These systems are more complicated, require more space and increase the investment costs.

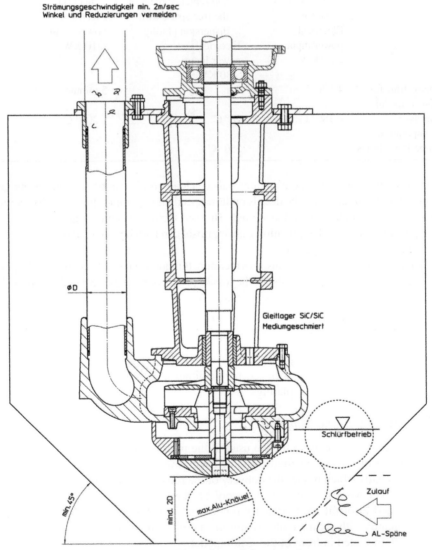

Graphic 7 Installation of the pump with cutting unit in a container

Another version of a single-stage centrifugal pump in a compact bloc construction is a type, equipped with a cutting unit, pre-built in front of the impeller (see graphic 7 [24]). The cutting unit is fixed on the extended pump shaft. Here, the cutting unit sits directly in front of the pump case, but outside the casing, however. As soon as the fluid is drawn in with the chips, the chips can be crushed by the cutting edge, before they enter the pump casing. The cutting edge crushes the steel wool balls and cuts the chips (see figure 80 and 81 [24]). The pump can thus work blockage freely through this and the good sucking characteristics remain unchanged (see figure 82 and 83 [24]).

Figure 80 Cutting unit with cutting plate

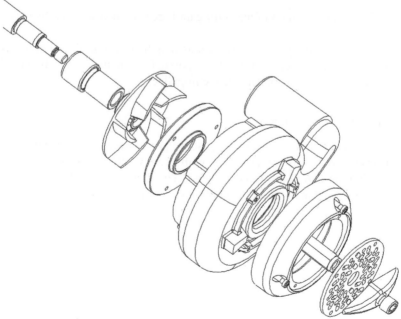

Figure 81 Pump with cutting unit

Figure 82 metal chips before crushing

Figure 83 metal chips after crushing

Cutting plate and the edges of the cutting unit are hardened and thereby very resistant to wear-out.

The holes in the cutting plate are shaped as long holes and designed in a way that the pump inside is protected from too large parts. Thus items, such as broken drills, cannot reach the pump and not damage the impeller.

4.5.2 Pump with Inducer

By different applications, pumps work in some processes with low negative suction heads. This can be the case by processes with negative pressure or if delivery fluids with a low boiling-point have to be pumped. The required NHSH value has then a borderline value, so that the risk of cavitation is very high.

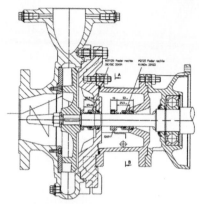

Graphic 8 Pump with Inducer [24]

Figure 84 Inducer

To minimize this risk and possibly exclude damages, an inducer-impeller can be assembled on the pump shaft, directly in front of the pump impeller (see graphic 8 [24]).

This "Header impeller" with a helical geometry increases the pressure at the impeller inlet as it were a "pre-pump". It minimizes the risk of cavitation and prevents the resulting damages from it (see figure 84 [24]). In addition, through the inducer, noise and energy losses are minimized.

5 Mobile and Stationary Damage Monitoring

Stationary systems for condition or damage monitoring are generally more complicated and more expensive than mobile systems. Criteria for the selection are: demanded frequency of the measuring signals, necessity of a trend analysis, required precision of the measurements, exposure to be expected and the expected risk of damage. Last not least, a costs-benefit analysis is decisive.

5.1 Mobile Damage Monitoring

If a pump is newly installed, it makes sense to monitor this unit for a given time. Important knowledge arises about this, whether the lay-out of the pump had been correctly done (type) and whether all technical requirements are fulfilled. The analysis of the pump readings and the noise level deliver important knowledge about the process flow and operating condition of the pump. It can also be found out whether the pump runs in the scheduled operating point.

If a new pump with different specification shall replace another pump, can be investigated by the monitoring whether the pump fulfills the full function of the old pump and whether the performance of the pump is sufficient. Thus the risk of a damage occuring later, can be minimized in an early stage. Provided that the results of the monitoring are satisfactory and the regular processes do not lead to any disruptive incident, the pump monitoring system can be dismantled again and installed on another pump.

The parameters, described in chapter 2, like pressure, temperature, speed, etc. can be measured with mobile gauges. Mobile compact devices are also available for measuring vibrations, condition of the ball bearings and for cavitation, even for evaluation and documentation. Much effort and costs can be economized by such mobile gauges since the mobile, short-time monitoring is sufficient for less critical systems and no stationary measuring system is necessary.

5.2 Stationary Condition Monitoring

Monitoring with a stationary system makes then sense, if endangering delivery media are in use or if a disturbed process flow of the pump system can cause very big damage. Hazardous materials or fluids can become risky by too high pressures or temperatures or lead to damages by leakage or overflowing of the delivery medium from the container. Often, the subsequent costs exceed even the costs for the real damage.

Critical processes, which require a very continuous delivery flow over a longer time period - at least in the 24 hour operation - can hardly resist without a stationary

Damages on Pumps and Systems. http://dx.doi.org/10.1016/B978-0-444-63366-8.00005-X

condition monitoring system. Strong fluctuations of the system parameters can be commanded by monitoring and controlling. Particularly in explosive areas, a monitoring system is very essential. It has to be distinguished between dust explosion and gas explosion. The pump- and plant components must comply with the valid EU instructions for the respective, explosive atmosphere. Here a so-called ATEX admittance is required. By very complex processes, in addition, the opportunity of the teleservice by a special company is obvious.

5.2.1 Structure of Stationary Condition Monitoring

For the exact measurement recording, it has to be paid attention on a proper installation of the individual components and the equipment being part of the system. Apart from the pump accounts also sensors, data lines, measuring box, PC and network.

Measuring errors appearing by wrong installation can have devastating consequences. Also applied software must be checked regularly.

For setting up of a monitoring concept, measurements must be carried through to investigate the limiting value. To this, specifically measurements at disturbances are carried out. The sequence of operations forms itself as follows:

- Measuring of the disturbances
- Storing the readings
- Analysis of the readings
- Specification of the limiting values
- Documentation in spreadsheet systems

The measurements in the process then require:

- Measuring of the process parameters
- Analysis of the readings
- Comparison of the values with the limiting values
- Identification of the limiting value exceedings
- Evaluation of the limiting value exceedings

If disturbances appear, there are different possibilities to reprocess the disturbing reports:

- Display on the screen with error message
- Display on the screen with error message and transmission to a primary terminal
- Display on the screen with error message and immediately reaction e.g. "pump stop".

The disturbing report can also be defined and transferred as an alarm report such as:

- Production stop
- Alarm transfer in the company's network
- Alarm by e-mail to offices outside
- Alarm by e-mail to servicing company

If an automated condition monitoring system shall be implemented, a regular, also automated analysis of the readings takes place. This analysis means that the identification of a disturbance through this results, that a measurement comparison with the deposited limiting values is carried out.

The following signal on the recognized disturbance according to a limiting value exceeding can be reprocessed in a very different way:

- Display at the monitor
- Ring tone
- Flashing signal lamp
- Turning the system off
- Check of other sensors, e.g. smoke detector, room temperature gas-analyses, etc.

The stationary condition monitoring offers a reliable measure for the early and quick damage limitation. For example, heavy and expensive damages can be prevented by switching off on time.

After implementation of a risk analysis and estimating the damage potential, it can be decided about the type and scale of the monitoring system.

5.3   Practical Examples of Vibration Measurement

The damage diagnosis by condition monitoring and vibration analysis can be provided online, or if a disruptive incident had the destruction of the pump as consequence, made by the analysis of the recorded readings. But by accurate condition monitoring, a damage can be avoided completely.

Every type of pump has its own typical characteristic vibration. These vibrations change by different operating states like cavitation, dry run, etc., but also by faulty components on pumps, such as mechanical seal.

So with the help of speed, a clear vibration "fingerprint" can be created for every pump. Through this, it is possible to make a reliable statement about the general condition of the pump. Speed sensor see figure 86 [12].

5.3.1   Construction of the Measuring Technique

Different sensors are shown in the following.

Figure 85 Acceleration sensor

Figure 86 Speed sensor [12]

Figure 87 Thermocouple

Figure 88 6 channel-measuring box

The measuring chain for every parameter to be measured is configured as follows:
Test point – sensor – cable – measurement amplifier – PC

The acceleration sensor (see figure 85 [12]) measures the impact sound in terms of to the vibration analysis, the thermocouple (see Figure 87 [12]) is used for the temperature measurement.

Figure 89 Telemetry-transmitter [12]

With the help of a telemetry system (see transmitter figure 89, and receiver figure 90 [12]) the readings can be transferred cableless. The number of channels with the signals to be transferred is limited on the possibilities of the utilizable radio frequencies. The processing of the readings takes place with a measuring box (see figure 88 [12]) near to the pump. From there, the digital data can be transmitted with Ethernet up to 100 meters of distance to the main terminal.

Figure 90 Telemetry-receiver [12]

5.3.2   Avoidance of Measuring Errors

A measuring-error can completely destroy the monitoring or measurement. The measurement has either a spread of more than 10% or it is completely wrong. Depending on the application, an inaccurate measurement is hardly to reprocess and often useless.

By a measurement, it has to be paid attention to the following things:

- Right selection of the sensors
- Right positioning of the sensors
- High measuring frequencies, otherwise not all vibrations can be measured
- Noticing the instruction from the sensor manufacturer on the right calibration
- Suitable time intervals - averaging
- Using measuring amplifier if necessary
- Taking into account tolerances/range of fluctuation of the parameters to be measured
- Taking into account environmental conditions like temperature, air humidity, etc.
- Cost relevance of the precision
- Precision: as exact as possible but as exact as necessary

If several parameters, which were measured wrongly or inaccurately, have influence on the calculation, this leads to error propagation. At the wrong end result, then it cannot be comprehended anymore, where the mistake comes from. Only by laborious computation and follow-up measuring, the result can be corrected.

5.3.3 Analysis of Measuring Signals

After comparison of the measurements with the limit value data, conclusions arise about the operating condition of the pump:

Speeds

Normal operation:	The speed of the pump is in the rated range.
Inlet end closed:	The speed is increased.
Increased air amount:	The speed is increased strongly.
Cavitation:	The speed is increased absolutely very strongly.

Temperature Measuring

Normal operation: A normal temperature profile by warming up. The ball bearing heats up itself due to the friction. Due to flowing water in the pump the temperature of the mechanical seal remains constant.

Pressure side closed: Due to friction of the impeller with the water, the water temperature increases, since no fresh water can flow in. The temperature of the ball bearing increases through the warming of the casing.

Cavitation: not sufficient refrigerant water in the case, thus temperature of the mechanical seal and the ball bearing increases.

Dry run: Quick increase of the temperature in the mechanical seal visible.

5.3.4 Damage Diagnosis with the Help of Vibration Analysis

As already described in chapter 3.4.5 (pump monitoring) by the vibration diagnosis, the measured speed is additionally considered. The acceleration- or the impact sound-sensor measures the impact sound-vibration of the pump. As a measurement the acceleration is measured in g (9.81 m/s^2).

The analysis of the present operating condition of the pumps is carried out by means of frequency analysis. Moreover, the readings are loaded and FFT spectra are calculated (**F**ast **F**ourier **T**ransformation). Further, the level values and the speed are calculated according to the data format.

For every type of disturbance (dry run, cavitation, closed valve) a combination of limiting values, limiting curves, increasing values, characteristics can be determined. This makes it possible to diagnose the exact type of disturbance.

For the trend analysis for the maintenance cycles of the pumps, the measured data are recorded. The data from the past are completed with the current ones and illustrated and analyzed together, e.g. in 3-D-format. A group of disturbances is signalized before the error occurs, if given sizes are changing.

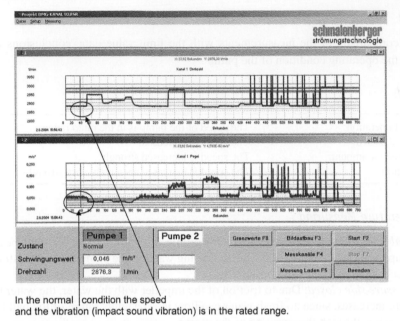

In the normal | condition the speed
and the vibration (impact sound vibration) is in the rated range.

Figure 91 Pump in the normal condition (vibration and speed) [24]

In the above diagram see figure 91 [24], the speed and the vibration are illustrated. The plain text display "normal" states, that pump No.1 is in the normal condition and pump No.2 is switched off.

This system is a self-learning system. By the first machine-operation, the actual values of the pumps to be monitored are measured and recorded. A characteristic is made for every pump (disturbances are feigned and adjusted), a rated range of the parameters (green range) and a disturbance range (red range) are defined, so that later, the system immediately diagnoses the condition of the pump, i.e. type and dimension of the disturbance.

If the inlet end is closed, vibration and speed are higher than the rated value.

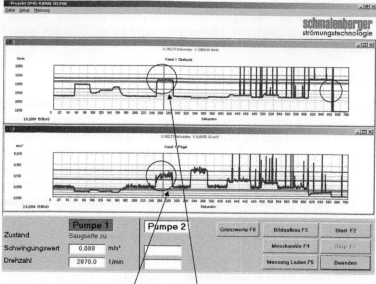

If the inlet is closed, vibration value and speed are higher than the nominal value.

Figure 92 Operating condition: Inlet end closed [24]

In the operating diagram (see figure 92 [24]), speed and vibration are also illustrated. The plain text display "inlet end closed" states, that pump No.1 is in a disturbance condition. The measurements for vibration and speed are too high, they are above the values for the normal condition.

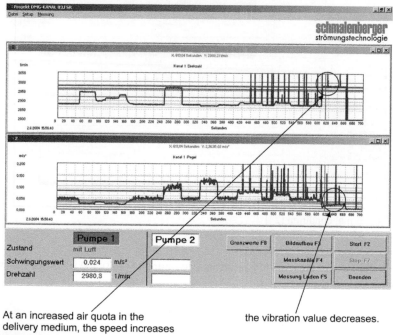

At an increased air quota in the   the vibration value decreases.
delivery medium, the speed increases

Figure 93 Operating state: increased air quota [24]

In the diagram "with air" (see figure 93 [24]) the plain text display "with air" states, that the pump is in a fault condition. The measurement for the vibration is too low, the value of speed is too high.

Telemetric Transmission

As already mentioned above, the readings can be transferred wireless with the help of a telemetry system. The sensors deliver analogous readings to the attached transmitting station, which is mounted near the pump. Before the telemetric transmission occurs, the analogous measuring signals are digitized and transferred in the MHz tape (megahertz) to the receiver. The telemetry receiver station is located at the measuring PC or at the PC-control station. The transferred digital readings are at first converted for the further processing and then written into a file or deposited in an allocated RAM memory. [12]

Impact Sound Measurements with Acceleration Sensors

In the following diagrams, the time courses of the values of the attached sensors with the different operating states or disturbances are shown:

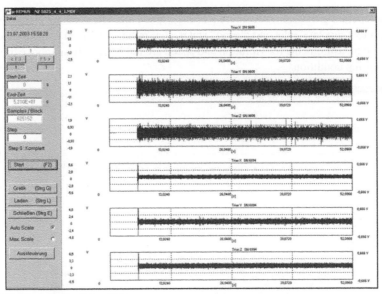

Figure 94 Normal condition [24]

The pump runs in the normal condition (see figure 94 [24]), the impact sound vibrations are constant and balanced.

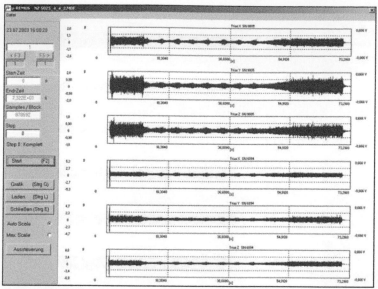

Figure 95 State with air [24]

The pump delivers air within the medium (fluid) (see figure 95 [24]), the impact sound vibrations are unbalanced and strongly fluctuating.

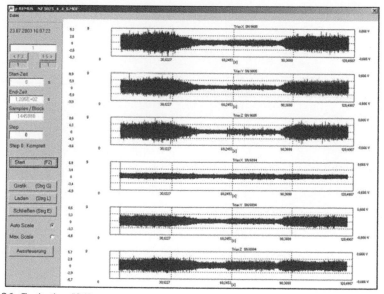

Figure 96 Cavitation [24]

The pump is cavitating (see figure 96 [24]). It comes to a local pressure lowering by over-speed and to vapor lock by evaporation of the delivery fluid. The impact sound vibrations are uneven, pulsating and with a high amplitude.

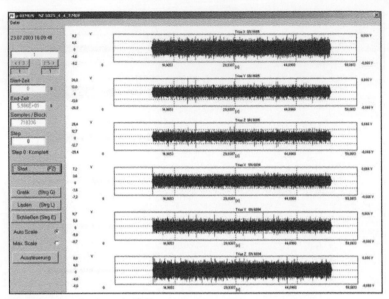

Figure 97 Pump runs backwards [24]

The pump runs backwards (see figure 97 [24]), it delivers fluid, but too little. The impact sound vibrations are relatively constant but with a high amplitude.

6 Advices for Planning and Conception of More Predictive Maintenance

A trouble-free operation of pump systems can be made possible by a future-oriented, intelligent maintenance strategy. Maintenance consists mainly of the basic measures servicing (maintenance), inspection, repairs and improvement in accordance with the Norm DIN 31051. The maintenance includes the operation of a pump system over the complete life cycle, from the first machine-operation up to the disposal.

While the maintenance includes basic measures like check, cleaning or also exchange of wear-out components, more extensive measures are carried out by the inspection. Data of the current state are compared with those of the nominal condition and possibly measuring of operation- pressure, temperature, electric current consumption, speed etc. are necessary.

By repair it is assumed, that the pump system is completely or partial out of order or there is a service-interrupting disturbance. Weak points or damages are cleared by this method. The improvement of maintenance means, increasing of the operational function of the system, so that interruptions or loss of productions are preferably avoided. On one hand this might mean the precautionary exchange of one or more components, or the realization of condition-oriented maintenance. With regard to the future, before damage enters, maintenance measures can be started by implementing continuous monitoring of the system. The measuring and recording of relevant parameters follows the analysis and evaluation and then the planning and implementation of measures.

Therefore, unforeseen disturbances and failures can be avoided, repairs can be minimized. Not but least, the saving of reserve pumps provides considerable economic advantages. About type and range of the monitoring system must be decided on the individual case, depending on application.

6.1 Monitoring of Pumps

Till now, the failure-oriented maintenance is still practiced. The pump is operated until the breakdown, only then, the worn out components are replaced. The time based maintenance contains precautionary, to replace the wear out relevant components regularly, even if they are not yet worn out.

Through condition-oriented, foresighted maintenance, standstill of systems, which means costs and loss of production, it is supposed to be avoided. Through careful work, carried out as a precaution and replacement of components, direct costs can be saved.

Damages on Pumps and Systems. http://dx.doi.org/10.1016/B978-0-444-63366-8.00006-1

6.2   Diagnosis Systems

Both by selective inspection and by the online monitoring, the process consists of the following single steps:

*   Measuring, recording, memorizing
*   Calculating, evaluating, reviewing
*   Illustrating, classifying, alarm diagnosing
*   Measure-planning and -implementation

As already described in chapter 2, the special system-/ pump parameters must be registered and evaluated to plan suitable measures. Pressure, temperature, speed, or electricity gauges can be installed as stationary indicating instruments without further functions, for the regular inspection in the system. Furthermore for regular inspection, mobile diagnostic systems are implemented, which absolutely have also analysis functions besides the measuring functions. Process parameters like pressure, temperature or flow rate as well as vibrations, speed or unbalanced mass, are recorded and evaluated.

With a mobile equipment, signal analyses, trend notes, till complete machinery diagnoses can be prepared.

Through appropriate software, measurement tasks can be programmed. During the measuring, a comparison of the measured values with predefined limiting values is carried out. After recording and evaluation, it comes to the preparation of diagnosis reports.

Stationary diagnosis systems for the online condition monitoring, make sense for complex systems. Such systems combine recording, measuring, regulating and controlling (EMSR) up to the documentation of the results in one device.

On the monitoring terminals, the current system parameters, disturbances, alarms or other notes are notified for the maintenance personnel. By special maintenance software, informations and suggestions can be indicated, such like maintenance measures up to spare part provision.

The monitoring, together with documentation of all maintenance measures, makes an almost trouble-free operation of the systems possible. For the calculation of the profitability, effort and costs must be compared to the costs for loss of production and resultant damages. Since such systems also record trends and cause with regard to the future to avoid damages, often the technological effort is obviously justified.

6.3   Data Transmission

If it is a new plant, the data transmission is integrated into the network of the entire system for the pump monitoring or will be connected to a SPS control unit. Professional bus (Profibus) and Industrial Ethernet systems are far common below the variety of the common networks. Also via wireless communication systems (Profinet), the data can be transmitted. However, the most important factor by the planning is, that the appropriate interfaces are available. Currently, important and far common systems are shown in table 16 [19]:

Table 16 Field-Bus Systems and Industrial Ethernet Systems

Industrial-Ethernet-Systems	Field-Bus-Systems
Ethernet/IP	Profibus
Profinet	CANopen
Ether CAT	Devicenet
SafetyNet	CC-Link
Powerlink	Modbus
Sercos	

To avoid, having to coordinate different systems with each other, it is advised, to use Profinet due to its universality and openness.

In the area of the motor drive, the Ethernet technology offers the possibility of exchanging the data in cycle time of the motor regulation, with the control unit in real time. Concerning the flexibility, Ethernet/IP and Profinet, are widespread in the drive technology.

6.4 Teleservice/Remote Maintenance

If once the servicing concept is integrated into an industrial network, the way to a teleservice- system is not very long. For example, Profinet permits to access to the data of a system via the web server, regardless of the location. If a remote diagnosis interface is available, the measuring signals of the sensors can be recorded and evaluated online. This creates the possibility of clearing smaller disturbances, without travelling expenses arising immediately. The complete diagnosis sequence of operations can so be managed, beginning with the recording of the measurements, the calculation and evaluation of the data.

The Ethernet technology therefore creates possibilities of assigning the monitoring and servicing tasks as an external service, by easily manageable costs.

6.5 Diagnosis and Maintenance as a Service Supply

To allocate maintenance and servicing externally, means nevertheless that the operator or owner of the plant knows his data for the pump system exactly. It is supposed that the critical areas or the system parameters being monitored, have to be known. Also previously it has to be cleared, if a regular inspection carried out in firm intervals is sufficient, or if the online monitoring makes sense.

Risks of failure and failure costs must be compared with the effort for the servicing. Staff costs and measuring equipment, arising in the house of the owner of the plant, are another criterion for the decision.

Table 17 Comparison: Inhouse Servicing or External Servicing

Criteria for Servicing/ Maintenance – Option	Maintenance Personnel of One's Own	External Maintenance by Service Providers
Trained specialist staff	+	++
Projectable fixed costs	+	++
Flexibility	++	+
The most modern measuring technique	+	+++
Risk appraisal	+	++
Reliable, periodic maintenance	+	++
Execution of small repairs	++	+

In general, external service providers have especially trained specialist staff and the measuring equipment corresponding to the state of the art. Both, full servicing and only individual modules offer the service providers. Whether only the monitoring or also complete repairs are assigned, this then must be decided by individual case. The conclusion of a maintenance agreement might be possibly a suitable solution for smaller enterprises. A comparison is shown in table 17 [T. Merkle].

7 Predictive Maintenance – Economic and Efficient

The economic feasibility of different measures orientates itself at the benefit and the costs arising out of the measures. However, decisive is the time period of the operation of the pump between going into operation up to the disposal. The cost pool for the investment is only by few pump systems more than 10% of the total costs, by most systems it's rather less. Costs for servicing, repair and energy predominate. Therefore a trouble-free energy optimized operation is very reasonable.

7.1 Best Efficiency Point

The optimal operating point adaptation is the simplest and most effective way of saving energy and costs. The difficulty often consists in investigating the system characteristic curve exactly, and in fixing the operating point consequently. In the optimal operating point, the pump has the best degree of efficiency, the optimal energy consumption, and is operated most gently, regarding the mechanical strain. The individual opportunities for the optimization were already described in chapter 3.1.3.

7.2 Energy Efficiency

Globally viewed, more than 20% of the world's electrical energy requirement is consumed by operating of pump systems. One aim is, to conserve resources and to counteract the energy shortage, on the other hand cost-saving by efficient pump operation opens great opportunities.

The efficiency of pump systems depends substantially on the kind of operation, which means adaptation of pressure and flow rate to the actual requirement. To be able to make a more precise statement about this, different types of costs of the system must be considered for the whole life cycle. An investigation of the VDMA (German association for machinery) showed, that the initial costs represent only about 10% of the life cycle costs, the energy costs, however, generate the greatest cost quota with more than 40%.

The use of energy saving motors can make a good contribution to the energy savings in the system. However, the speed control by frequency converters enables the three to fivefold of energy savings. However, the greatest reduction potential exists in the hydraulic optimization of the whole pump system.

Damages on Pumps and Systems. http://dx.doi.org/10.1016/B978-0-444-63366-8.00007-3

For a more detailed estimation of the reduction potential, an analysis of the entire system must be carried out.

According to evaluation of the ZVEI (central association of electrical engineering and electronics industry in Germany) the energy reduction potential with electromotive driven systems constitutes as follows:

- energy optimized engines: 10%
- electronic speed control: 30%
- mechanical system optimization: 60%

Energy Optimized Motors

The attention increasingly turns to the efficiency or to the system effectiveness of the electrical drives. In accordance with the new EU-regulation 60034-30:2009, since June 16 th, 2011, the motors have not been evaluated any more after the EFF- classification, but subdivided to IE categories (International Efficiency).

This norm stipulates minimum efficiency factors of IE2 for three-phase motors Time schedule see table 18 [35].

Table 18 Time Schedule with Stages of the EU-Regulation for the Increase of Effectiveness at Electrical Motors [35]

Stages of the EU-Regulation No. 640/2009 (50 Hz) of the European Union	Requirements	Notes
Stage 1: **As of 16th June 2011**	The motors must fulfill the efficiency factors of IE2	IE2/high efficiency factor, comparable with EFF1 (European CEMEP agreement)
Stage 2: **As of 1st January 2015**	Motors with a nominal power of **7.5 - 375 kW** have to fulfill the efficiency class IE 3, or if they are equipped with a speed regulated drive, class IE2	IE3/premium- efficiency, derived from the class IE2 with ca.15% lower losses
Stage 3: **As of 1st January 2017**	Motors with a nominal power of **0.75 - 375 kW** have to fulfill the efficiency class IE 3, or if they are equipped with a speed regulated drive, class IE2	

The development in the direction of the energy efficiency of motors will go on continuously. The new EU-regulation states further, that three-phase motors of 7.5 kW up to 375 kW must either correspond to the higher efficiency class IE3 or be equipped with a frequency converter in the IE2 version from January 1st, 2015.

From January 1st, 2017, this regulation also concerns motors from 0.75 kW up to 7.5 kW.

These requirements technologically are released by different variants:

- More copper in the motors
- Motors with an integrated frequency converter
- Synchronous motors with permanent magnet

The use of energy optimized motors does not only make sense for energy saving reasons. For the operation of pumps which are often working for several decades, results also a considerable cost saving. Depending on pump size and operation hours per year, such pumps will have amortized themselves already after a few years.

Since the more effective motors develop less waste heat, in many cases no fan wheel is necessary. The optimization of the electrical winding also leads into improvement of the efficiency factor.

Special ball-/ antifriction bearings (Low-torque antifriction bearings) provide a 40-50% lower friction torque. The lower friction of the bearing enables often the choice of a smaller electromotor. An improvement of the efficiency factor by 1.5% seems to be realistic. A high efficiency motor is shown in figure 98 [30].

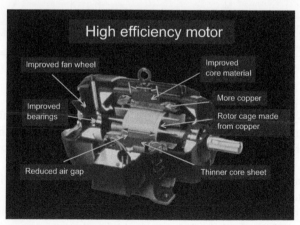

Figure 98 Opportunities of optimization at the electrical drive [30]

Based on the fact, that pumps in the future are equipped with much more effective motors, a further potential of optimization can be realized constructively.

Investigations showed, that regarding the pump hydraulics there is still absolutely potential. By improvement of the impeller surface, back plate and spiral casings, the efficiency factor can still be increased by some percent. Optimization of clearance and type of relieving (holes, etc.) make a further contribution.

Well balanced impellers causes besides less noise level also a lower energy consumption.

The reduction of the wall thicknesses at the impellers causes a weight saving. By the low weight, the moment of inertia - Jw - will be reduced. Especially by frequent switching on/off, and as well at speed modifications, this brings saving of energy too.

By mechanical seals, the friction resistance can be reduced. This will be possible by smoothing of the touch surfaces and positioning a thin lubricating film between the slide surfaces. A raised friction resistance is converted into heat loss and reduces the efficiency. Diamond-coated mechanical seals are even better.

The expected "sum" of this energy savings by the component "pump" could yield +/− 10% improvement of the efficiency factor.

After the analysis of the component "pump", a complete system analysis of the pump system has to be carried out as shown in graphic 9 [30]:

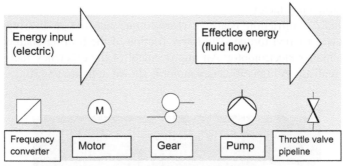

Graphic 9 Single components of a pump system - optional [30]

Thus pump systems can often be very complex and sometimes containing very different components, by system analysis, each component must be examined separately. Each individual efficiency factor of a component affects the total pump efficiency significantly.

An optimal adjustment to the operating point can be realized by trimming of the impeller diameter and the use of frequency converters.

The process optimization (cascade system, exact definition of total head and delivery rate) offers - independently of the pump - a high reduction potential.

The pump power output or efficiency in comparison with the necessary electrical performance is represented in the graphic 10 [30]. Here, the effective output was set at 100%. Comparing the two energy flow diagrams, it is clearly to identify, that the main potential of savings is the process of the power unit.

By conventional throttle control, the 2.85-fold of the pump power output must be spent as electrical energy. The electronic speed drive demands however, only 1.6-fold.

Less energy consumption of the complete system could absolutely rise the system effectiveness about 30-35%. The improvement can be caused: by additional measures like an optimal adaptation of the operating point, optimization of piping and valves, speed control by frequency converters, as well as process optimization of the system.

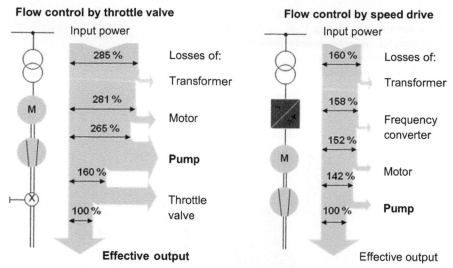

Graphic 10 Comparison of throttle control with speed drive [30]

7.3 Life Cycle Costs

To be able to analyze the costs of pumps more precisely, it is necessary to carry through a total cost analysis. This consideration includes the calculation of the so-called "life cycle costs".

These aspects of TCO (Total Cost of Ownership) and LCC (Life Cycle Costs) are situated more in the focus on the analysis of the service life of a pump system.

The purpose of many operators is, buying pump systems where the complete costs can be exactly defined. Here, all costs like purchasing costs, operating costs, up to costs for disposal are taken into account and are exactly defined over a time period of 20–30 years.

In these life cycle costs (LCC), the costs for maintenance and repairs, and especially the availability of the system respectively contains the costs for the failure by standstill of the system.

This means, in the context of the TCO consideration, the value for the higher service life has influence directly on the economy calculation.

The multi investment into high-quality pumps with special coatings, energy-saving motors or intelligent controls, in most cases, it is much more economic, than the renunciation of it. The costs for repairs and failure of the pump systems are much higher.

The life cycle costs [LCC - Life Cycle Costs] consist in the sum as follows:

$$LCC = \text{sum of: } C_{ic} + C_{in} + C_e + C_o + C_m + C_s + C_{env} + C_d$$

Where:

C_{ic} = initial costs (investment)
C_{in} = installation and putting into operation
C_e = energy costs

C_o = operating costs (running costs)
C_m = downtime, production loss
C_s = costs for maintenance and repairs
C_{env} = environment costs
C_d = closure and disposal

The main cost factors arise as the result:

Main cost factors

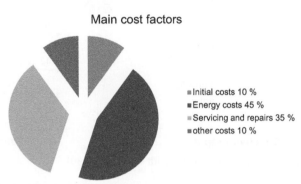

- Initial costs 10 %
- Energy costs 45 %
- Servicing and repairs 35 %
- other costs 10 %

Diagram 25 Sectioning of costs at pumps

Initial costs (investment), maintenance, energy costs, other costs

By analysis of the cost quotas it is inestimable that the main attention must be focused on maintenance and energy consumption (see diagram 25). Solutions to this are: Wear out protection for the reduction of maintenance and repair, optimization of the energy consumption by frequency converters, hydraulic optimization of the system as well as energy optimized motors.

7.3.1 Initial Costs

Since the initial costs only make an impact of about 10% at the LCC consideration, the purchase price is less relevant than the other costs. If so an economical pump fails after 2 years and must be replaced, no costs are saved in comparison with a pump which is working for 15 years. The economy can be calculated with already available LCC software of different suppliers to carry out a comparison of individual pump varieties. So the purchasing price of a pump must always be compared to the other costs, especially costs for energy and maintenance.

7.3.2 Energy Costs

In comparison with the initial costs which once arise as a fixed amount (incl. discounting), the energy costs are very dynamic.

Although, energy costs are subject to fluctuations also, the tendency was and is, however, always increasing. At a quota of the energy costs of approx. 45%, this is the most critical factor of LCC, because the rate of price increases is only very difficult to anticipate.

Saving of energy by system optimization, using energy-saving motors or electrical speed control, is therefore economic in every case. The payback periods last here from some months up to some few years.

This is represented in detail in the following at two concrete examples (see table 19 [5]):

Example 1 Process Pumps:

Table 19 Costs- and CO_2 - Savings for a Pump System [5].

Before the Optimization	After the Optimization
Control: Bypass + throttle valve	Frequency converter
Oversized pump, 18,5 kW	Energy efficient pump, 1,5 kW
Electricity consumption: 57 000 kWh	Electricity consumption: 5 590 kWh
	Investment: 3 800,- €
Costs of electricity: 6840,- €	Costs of electricity: 670,- €
Reduction of the electricity consumption: 51 410 kWh/Jahr	
Cost reduction: 6 170,- €/year (by cost of electricity of 12 cent/kWh)	
CO_2 - reduction: 33 t/year (by German electricity mix: 633 g CO_2/kWh)	
Payback period: < 1 Jahr	

Example 2 Process Pumps (27 Pump Systems Approx. 2 500 Pumps, see table 20 [5]):

Table 20 Costs- and CO_2 - Savings at More Pump Systems [5].

Before the Optimization	After the Optimization
Control: Bypass + throttle valve	Frequency converter
Oversized pumps	Energy efficient pumps, IE2-motors
Electricity consumption: 2 305 000 kWh	Electricity consumption: 1 678 530 kWh
	Investment: 111 530,- €
Costs of electricity: 276 600,- €	Costs of electricity: 201 424,- €
Reduction of the electricity consumption: 626 470 kWh/Jahr	
Cost reduction: 75 180,- €/year (at cost of electricity of 12 cent/kWh)	
CO_2 - reduction: 400 t/year (at german electricity mix: 633 g CO_2/kWh)	
Payback period: ca. 1,5 years	

By the replacement of the throttle control or the bypass control through a speed control with frequency converter, the efficiency of the pumps can be increased by some per cent.

7.3.3 Servicing and Repairs

Qualitatively good pumps have in many cases higher prices, however usually the pumps are maintenance-free and less vulnerable to failure.

However, periodic maintenance in accordance with the given maintenance schedule helps to minimize the total costs. Especially for critical or complex systems, a foresighted maintenance is highly recommended. Suitable condition monitoring systems (C.M.) are already available on the market. Disturbances, which means costs, can at long last be avoided. Also here, the economy or the period of amortization can be calculated by means of LCC software.

7.3.4   Other Costs

Another tenth of the total costs can be summarized under the factor "other costs". These are costs for adjustment/putting into operation, operating and environment costs as well as costs for production loss during disruptions.

An installed condition monitoring system, which helps avoiding disturbances and subsequent loss of production, has amortized itself in most cases very fast. Costs for closure and disposal do not have an effect so much at a long life time of a pump. But indeed on a short life time.

7.3.5   Software for LCC Calculation

Different software systems for the LCC calculation are already available on the market.

- Ball bearings: http://bearinx-online-easy-friction.schaeffler.com
- Systems: www.world-class-manufacturing.com
- Pump systems: www.system-energieeffizienz.de
- Pumps/electrical drives: www.zvei.org/Lebenszykluskosten

The life cycle cost calculator for pumps makes it possible, also to compare the systems with different load profiles and capacity profiles. In the program, different operation parameters can be entered for the calculation of different pump systems. The calculation result is a cost value, which allows to compare the life cycle costs of the systems for selecting the most favourable. In most cases, not the system with the lowest investment costs is the most economical one, but the system with the lowest energy- and maintenance costs will be the most favourable.

7.3.6   Summarizing Consideration the LCC

Costs for maintenance and energy represent together approx. 80% of the total costs.

Low-maintenance pump systems, which usually are more expensive in the acquisition but comply with higher quality standards are certainly quite worthwhile. Implementation of energy-saving measures, especially by pumps in the continuous operation is part of the most effective system optimization.

The costs for energy and material will definitely increase within the next years. On one hand, the energy costs are incorporated directly into the costs for the operation of a pump system per running hour or per annum. But since material processing, for example producing a cast case or the copper winding of an electromotor is energy consuming too, increasing energy costs also have an effect on the manufacturing costs, and therefore just multiple negative.

Moreover by other products already regarded as the parameter "pay back time", will be paid attention. This means the time period, in which an improved pump system amortizes itself. Only due to growing material shortage, will the material prices increase. The costs which are dependent on the energy costs, like extraction, transportation and processing have to be added.

Recyclable products with re-usable materials, will receive more significance. The more recyclable a pump is, the higher their residual value will be at the time of the disposal. This means by purchasing a pump, the investment costs will be viewed more under the aspect of the recycling value.

7.4 Cost Increase and Material Shortage

Damages and wear out are always related to material consumption, costs and lastly energy consumption. Therefore, that's why it is important to have the whole consideration. Increasingly, materials which are reusable will be used for the products. Energy costs will increasingly determine the design of the products.

7.4.1 Rise of Energy Costs

Within the last few years the energy costs have risen continuously. The stagnating or relieving crude oil production indicates a shortage of the oil reserves. The prices for energy increased extraordinary to never existing amounts in 2008, a year with a very high economic growth. The price of crude oil was at U.S. $140 per barrel in June 2008. After a decrease in the year 2009, the peak almost was reached again in 2011 (see diagram 26).

Scientist of the union "Association for the study of Peak Oil" expects a strongly declining crude oil production in the future. Therefore, the prices of energy will rise continuously and affect the production costs of pumps directly (see diagram 27).

7.4.2 Costs of Materials

The costs of material represent with more than 45% in the producing industry still in front of the personnel expenditures, the largest cost block bloc. While the labour productivity could be increased around the factor 3.5 since the year 1960, the development of the material productivity remained far back (factor 2). Studies have shown, that a 20 per cent increase of the material efficiency seems practicable up to the year 2015 (see diagram 28 [4]).

Since the costs of material are immediately bound up with the costs for transportation, energy and disposal, here, the potential for savings at an improvement on the material efficiency is very high.

Using material efficiently means both, to minimize the waste and offcut during the manufacturing of the products, and to optimize the completed product with regard to the material usage.

Modern computer-simulation tools offer here various possibilities to reduce wall thicknesses.

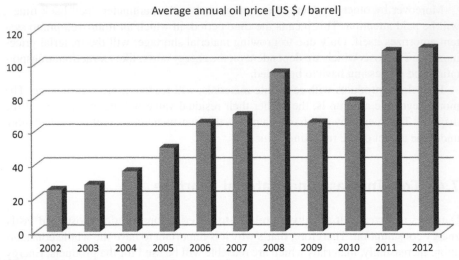

Diagram 26 Average annual oil price in U.S. $ per barrel

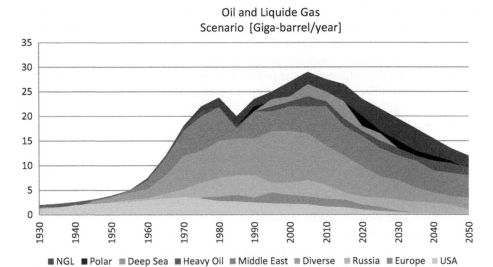

Diagram 27 Worldwide decreasing delivery of oil and liquefied gases, 1930-2050

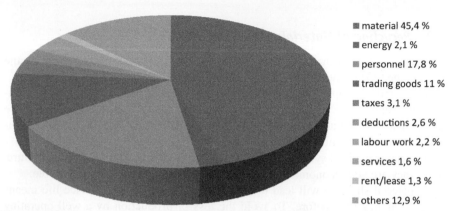

Diagram 28 Cost structure in the producing industry, [www.demea.de]

Production methods and product design must be adapted to the optimized material deployment.

In comparison with the costs for staff and energy, the costs for material have risen continuously within the last few years. On the one hand there is potential for savings,

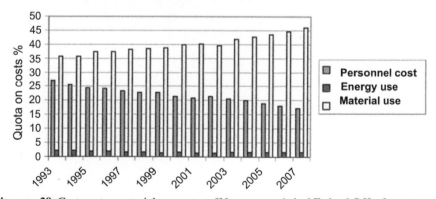

Diagram 29 Cost quotas material, energy, staff [source: statistical Federal Office]

on the other hand it is very important to organize the material usage in a way, that the products or pumps are able to work relatively long enough without breakdown and wear-out (see diagram 29 [10]).

However, to minimize damages to pumps, high-quality wear-out resistant materials are recommended. Having a view to the life time of the pump, where the initial costs represent the smaller quota, this clarifies that the investment into a high-quality, long-lasting and low-maintenance pump is worthwhile.

7.4.3   Recycling of Materials

Increasing energy and material prices will in future more and more determine the mode and the type of the pumps and systems.

Consequently, pump components made of the metals like steel, copper and aluminum will have to be long-lasting and recyclable. After the disposal, they will be carefully dismantled, put in pieces into their components and recycled. Even the components of the electro-motors will also be put into pieces and then reused.

Components made of synthetic/plastics material will only be competitive in future, if either they are very economically produced and disposable or fully recyclable.

The material shortage will lead to the fact that an increase of service life means more cash money than before. To avoid the loss of production by a well operating pump system will lead to even higher cost savings than before.

Used Formula Signs and Units

H	Total head
H_{stat}	static total head quota (plant)
H_{dyn}	dynamic total head quota (plant)
J_W	Moment of inertia
η	Efficiency factor
P	Power output
Q	Capacity (delivery rate)
$p1$	Pressure at the inlet side (suction)
p_{amb}	Air pressure
p_v	Vapour pressure
V_1	Flow speed at the inlet side (suction)
z_s	Geodesic height related to the datum plane
ρ	Density of the delivery fluid
g	Local acceleration of gravity

Units

A	Amperage
1/min	Rotation speed
rpm	Rounds per minute
bar	Pressure
°C	Temperature
f	Frequency (Hz)
HV	Harden in Vickers
K	Unit for temperature difference
kg/m^3	Solid matter quota
kg/l	Solid matter quota
kW	Performance / motor output
mg/l	Cl-Ion quota (Chloride)
m/s	Flow speed
m/s^2	Gravity ($9,81$ m/s^2)
N/mm^2	Tensile stress (Elastic modulus)
µm	Layer thickness
m^3/h	Flow / flow rate
V	Voltage

Bibliography

[1] Schmalenberger GmbH + Co. KG, internal papers, Balasz-Klein I. Steigerung der Energie-Effizienz von Kreiselpumpen – Untersuchungen zur hydraulischen und antriebstechnischen Optimierung. Bachelor-Thesis. Hochschule Reutlingen; 2010.

[2] Bohl W, Elmendorf W. Technische Strömungslehre. Würzburg: Vogel-Verlag; 2008.

[3] Bosch Rexroth AG. Rexroth Praxisseminar - Elektrische Antriebe 2005. Lohr, 2005.

[4] demea; Basisinformationen, warum ist Materialeffizienz wichtig. Berlin: Deutsche Material-effizienzagentur (demea); 2011. www.demea.de.

[5] dena; Erfolgsbilanz bei Pumpensystemen: Energieeffizienz lohnt sich. Berlin: Deutsche Energieagentur; 2007.

[6] dena; Contracting, Betriebsführungs-Contracting. Berlin: de, Deutsche Energieagentur; 2011. www.energieeffizienz-im-sevice.

[7] DIN 31051; Grundlagen der Instandhaltung, Deutsche Norm. Berlin: Deutsches Institut für Normung; 2003.

[8] DIN 50902; Korrosionsschutzschichten und ihre Herstellung, Deutsche Norm. Berlin: Deutsches Institut für Normung; 1994.

[9] Edelstahl Rostfrei - Eigenschaften; Merkblatt 821, Informationsstelle Edelstahl Rostfrei. Düsseldorf 2006.

[10] Energieeffizienz. Ministerium für Umwelt, Klima und Energiewirtschaft Baden Württemberg. de, Stuttgart; 2011. www.umweltschutz-bw.

[11] Kalenborn; Verschleißschutz für Anlagenkomponenten und Rohrleitungen, Kalenborn Kalprotect Dr. Mauritz GmbH & Co.KG, Vettelschoss.

[12] Hastrich HP. Schäden an Pumpen und Pumpensystemen. Esslingen: Seminar-Manuskript Technische Akademie Esslingen; 2010.

[13] Henzelmann T, Büchele R. Der Beitrag des Maschinen - und Anlagenbaus zur Energieeffizienz; Roland Berger Strategy Consultants. Frankfurt a.M: Studie im Auftrag des VDMA; 2009.

[14] Isecke B, Mietz J. Vermeidung der Korrosion nichtrostender Edelstähle in chlorid-belasteter Schwimmhallen-Atmosphäre, Dokumentation 882. Düsseldorf: Informationsstelle Edelstahl Rostfrei; 2001.

[15] Lenze; Dezentrale Antriebstechnik. Hamelin: Lenze Automation GmbH; 2011.

[16] Liedtke D. Nitrieren und Nitrocarburieren, Seminar-Manuskript. Esslingen: Technische Akademie Esslingen; 2005.

[17] Merkle T, Reuschel A, Schmauder S. Hart im Nehmen - Prozesssicherheit durch Verschleißschutz, Verfahrenstechnik. Mainz: Vereinigte Fachverlage; 2010.

[18] Merkle T. Verfahren und Vorrichtung zur Überwachung von Pumpenanlagen. München: Patentschrift DE 10 2004 028 643; 2005.

[19] Moxa Europe GmbH. Aus der Ferne überwachen – Echtzeit-Statusüberwachung im Ethernet-Netzwerk, messtec drives Automation. Weinheim: GIT-Verlag; 2011.

[20] Piesslinger-Schwaiger S, Zahel H. Höhere Korrosionsbeständigkeit von Edelstahl durch Polinox Protect und Polinox Protect TC. München: Poligrat AG; 2011.

[21] Process-Seminar; Pumpenseminar, Störungsfrüherkennung. Würzburg: Vogel-Verlag; 2005.

[22] Prognos AG. Energieeffizienz in der Industrie – eine makroskopische Analyse der Effizienzentwicklung unter besonderer Berücksichtigung der Rolle des Maschinen- und Anlagenbaus. Berlin: Studie im Auftrag des VDMA; 2009.

[23] Reuschel A, Merkle T, Schmauder S. Dem Verschleiß auf der Spur; Process 3–10. Würzburg: Vogel-Verlag; 2010.

[24] Schmalenberger GmbH + Co. KG; interne Schriften und Berichte 2003 – 2011.

[25] Sigloch H. Technische Fluidmechanik. Berlin: Springer-Verlag; 2008.

[26] Sigloch H. Strömungsmaschinen. München: Carl Hanser-Verlag; 2009.

[27] Siemens AG. Energieoptimierung für Industrie und Infrastruktur. Erlangen, 2008.

[28] Solid Works; Flow Simulation Tutorial, solidpro GmbH. Seligenstadt, 2011.

[29] Solid Works; Sieben Kerntechnologien der Flow Simulation, solidpro GmbH. Seligenstadt, 2011.

[30] Steins D. Betriebskosten senken mit Energiesparmotoren. Düsseldorf: Deutsches Kupferinstitut; 2011.

[31] Todorov V. Neue Materialien zur Erhöhung der Verschleiß - und Korrosionsfestigkeit von Kreiselpumpen, Studienarbeit, Institut für Materialprüfung, Werkstoffkunde und Festigkeitslehre. Stuttgart: Universität Stuttgart; 2005.

[32] VDI-Richtlinien. VDI 3822, Schadensanalyse; Grundlagen und Durchführung einer Schadensanalyse. Düsseldorf: VDI; 2010.

[33] VDI-Richtlinien. VDI 3832, Körperschallmessungen zur Zustandsbeurteilung von Wälzlagern an Maschinen und Anlagen. Düsseldorf: VDI; 2007.

[34] VDMA. Der Beitrag des Maschinen - und Anlagenbaus zur Energieeffizienz. Frankfurt a.M 2009.

[35] VDMA. Pumpen Lebens-Zyklus-Kosten. Frankfurt a.M 2003.

[36] Verschleißbeständige weiße Gusseisenwerkstoffe – Zentrale für Gussverwendung, Düsseldorf.

[37] Verordnung (EG) Nr.640/2009 der Kommission zur Durchführung der Richtlinie 2005/32/ EG des Europäischen Parlaments und des Rates im Hinblick auf die Festlegung von Anforderungen an die umweltgerechte Gestaltung von Elektromotoren. Brüssel 2009.

[38] Volz M. Nicht alle Industrial-Ethernet-Systeme werden sich am Markt durchsetzen, messtec drives Automation. Weinheim: GIT-Verlag; 2011.

[39] Vortmann C. Untersuchungen zur Thermodynamik des Phasenübergangs bei der numerischen Berechnung kavitierender Düsenströmungen. Karlsruhe: Universität Karlsruhe; 2001.

[40] Wagner W. Rohrleitungstechnik. Würzburg: Vogel-Verlag; 2008.

[41] Wagner W. Kreiselpumpen und Kreiselpumpenanlagen. Würzburg: Vogel-Verlag; 2009.

[42] Wärmebehandlung von Stahl – Nitrieren und Nitrocarburieren; Merkblatt 447. Düsseldorf: Stahl-Informations-Zentrum; 2005.

[43] EagleBurgmann Germany GmbH & Co. KG, Werksbüro Stuttgart, 2014.

[44] Kavitation an einem Tragflügel; Universität Stuttgart Institut für Strömungsmechanik und Hydraulische Strömungsmaschinen, Stuttgart.

Index

Printed and bound by CPI Group (UK) Ltd, Croydon, CR0 4YY

03/10/2024

01040427-0015